MODELING MORPHODYNAMIC EVOLUTION IN ALLUVIAL ESTUARIES

Modeling morphodynamic evolution in alluvial estuaries

DISSERTATION

Submitted in fulfillment of the requirements of
the Board for Doctorates of Delft University of Technology
and of
the Academic Board of the UNESCO-IHE Institute for Water Education
for the Degree of DOCTOR
to be defended in public
on Wednesday, 26 May 2010 at 10.00 hours
in Delft, the Netherlands

by

MICK VAN DER WEGEN
born in Amersfoort, the Netherlands

Master of Science in Civil Engineering,
Delft University of Technology, the Netherlands

This disseration has been approved by the supervisor:
Prof. dr. ir. J.A. Roelvink

Members of the Awarding Committee:
Chairman   Rector Magnificus, TU Delft
Prof. dr. A. Szöllösi-Nagy   Vice-Chairman, Rector UNESCO-IHE
Prof. dr. ir. J.A. Roelvink   UNESCO-IHE/TU Delft, Supervisor
Prof. dr. ir. H.H.G. Savenije   TU Delft, the Netherlands
Prof. C.T. Friedrichs, PhD, BA   Virginia Institute of Marine Science (VIMS),
Prof. dr. H.E. de Swart   Utrecht University, the Netherlands
Prof. dr. ir. M.J.F. Stive   TU Delft, the Netherlands
Dr. Z.B. Wang   TU Delft, the Netherlands
Prof. S. Uhlenbrook, PhD, MSc   UNESCO-IHE, Vrije Universiteit, TU Delft, the
                                    Netherlands

CRC Press/Balkema is an imprint of the Taylor & Francis Group, an informa business

Published by:
CRC Press/Balkema
PO Box 447, 2300 AK Leiden, the Netherlands
e-mail: Pub.NL@taylorandfrancis.com
www.crcpress.com - www.taylorandfrancis.co.uk - www.ba.balkema.nl

Cover image: Computed morphodynamic evolution in a 320 km long tidal
embayment with bank erosion after 0, 800 and 6400 years.

ISBN 978-0-415-59274-1 (Taylor & Francis Group)

Turangalîla - Olivier Messiaen

Abstract

Estuaries are valuable areas. Tidal flats, tidally varying water levels and the salt-fresh water interface are characteristics that form a unique ecological environment, which is often of international importance. Local fishery, aquaculture and tourism are important sectors that profit from the estuarine richness. Additionally, numerous ports are situated along estuaries, forming the logistical link between ocean transport and the hinterland. It is of utmost importance to understand estuarine processes so that impact of human interference (like dredging) and long-term changes (like sea level rise or a changing discharge regime) can be estimated and evaluated.

The main objective of this research is to investigate the governing processes and characteristics that drive morphodynamic evolution in alluvial estuaries by means of application of a process-based numerical model (Delft3D). The Delft3D model numerically solves the shallow water equations, which result in a detailed water level and velocity field at any moment and at any place in the model domain. This information is used to predict sediment transports and the resulting bathymetric evolution.

The research starts from a highly schematized configuration (a rectangular tidal basin filled with uniform sand and forced by a semi-diurnal tide) to study characteristic morphodynamic evolution over millennia. In subsequent chapters more processes are added, autonomous behavior is restricted and the timescale is reduced, to finally end up with a model configuration that matches the geometry of a real alluvial estuary and a 'comprehensive' timescale that allows model validation with measured morphological developments. The case studies used for model validation are the Western Scheldt estuary in the Netherlands and San Pablo Bay in California, USA.

The models described in the current research did not reach a state of equilibrium in the strict sense. During the simulations tide residual transports and morphodynamic evolution remain continuously present despite the constant forcing conditions. Nevertheless, model results show a decay in (the rates of) energy dissipation and tide residual transports. Model results correspond with the empirically derived 'equilibrium' relationship between the tidal prism (P) and the cross-sectional area (A), which is seemingly constant over decades. In contrast to exponentially converging basins often observed in reality, the model results suggest a stable geometry with a linearly converging cross-sectional area.

The main time scales observed are related to
1) pattern development (a balance between the adaptation length scale of the flow and the adaptation length scale of the bed topography),
2) the longitudinal profile (mainly a function of the development of overtides),
3) sediment supply to the basin (by bank erosion or river supply) and
4) sea level rise and land subsidence processes.

The geometry of a tidal basin has significant impact on its morphodynamic development. By starting from a flat bed and imposing the Western Scheldt geometry, model results show development of the channel-shoal pattern in accordance with the observed bathymetry. In case of the San Pablo Bay model, variations of the model parameter settings have limited qualitative effect on decadal sedimentation and erosion patterns. This suggests that the geometry itself plays an important role in the observed bathymetric developments.

Samenvatting

Estuaria zijn waardevolle gebieden. Het intergetijde gebied en de zout-zoet water interactie zijn karakteristiek voor een unieke ecologische omgeving die vaak van internationaal belang is. Lokale visserij, aquacultuur en toerisme zijn belangrijke sectoren die profiteren van de estuarine rijkdom. Daarbij vormen de havens in het gebied een belangrijke logistieke link tussen maritiem transport en het achterland. Het is van groot belang om estuarine processen te begrijpen zodat de gevolgen van menselijk ingrijpen (zoals baggerwerkzaamheden) en lange-termijn veranderingen (zoals zeespiegelstijging) kunnen worden ingeschat en beoordeeld.

Het voornaamste doel van dit onderzoek is om de dominante processen en karakteristieken te onderzoeken die de drijfveer zijn van morfodynamische ontwikkeling in alluviale estuaria door toepassing van een proces-gebaseerd numeriek model. Het Delft3D model lost op numerieke wijze de ondiep water vergelijkingen op, wat resulteert in waterstanden en stroomsnelheden op elk moment en op elke plaats in het model domein. Deze gegevens worden gebruikt om het sediment transport en de resulterende bodem ontwikkeling te voorspellen.

De studie begint met een grove model schematisering (een rechthoekig getij bekken met uniforme zand verdeling geforceerd door een halfdaags getij) die de karakteristieke morphodynamische ontwikkeling over millennia onderzoekt. In opeenvolgende hoofdstukken worden meer processen toegevoegd, wordt autonome ontwikkeling beperkt en wordt de tijdschaal kleiner gemaakt, zodat de studie uiteindelijk afsluit met een model configuratie die de geometrie van een echt estuarium voorstelt met een 'beperkte' tijdschaal, zodat model validatie uitgevoerd kan worden op basis van metingen. Case studies voor model validatie zijn het Westerschelde estuarium in Nederland en San Pablo Bay in Californie.

De in deze studie beschreven modellen leiden niet tot een staat van evenwicht in de strikte zin van het woord. Getij residuele transporten en morphodynamische ontwikkeling blijven continu aanwezig ondanks de constante forcering. Daarbij laten de modelresultaten een afname zien in termen van (de verandering in) energie dissipatie en getij residuele transporten. De model resultaten komen overeen met empirisch verkregen evenwichtsvergelijkingen tussen het getij prisma (P) en doorsnede oppervlakte (A), schijnbaar constant over een tijdschaal van decennia.

De voornaamste geobserveerde tijdschalen zijn gerelateerd aan
1) plaat-geul ontwikkeling (een evenwicht tussen de aanpassingslengte van de stroming en de aanpassingslengte van de bodemtopografie),
2) het langsprofiel (voornamelijk een functie van de ontwikkeling van hogere harmonische getij golven),
3) sediment aanvoer naar het getij bekken (door bankerosie of riviertoevoer) en
4) zeespiegelstijging en dalingsprocessen van land.

De geometrie speelt een significante rol in de morphodynamische ontwikkeling van een getijbekken. Als het model begint met een vlakke bodem en de geometrie van de Westerschelde, ontwikkelt zich en plaat-geul patroon dat in overeenstemming is met de waargenomen morfologie. In het geval van de San Pablo Bay case studie leiden variaties in model parameter instellingen maar tot een beperkt kwalitatief effect op de ontwikkeling van het erosie en sedimentatie patroon over tientallen jaren. Dit suggereert dat de geometrie van een bekken een belangrijke rol speelt in de morfologie.

Acknowledgements

About 100 years ago Osborne Reynolds carried out laboratory tests with a physical scale model of water and sand covering a similar topic as described in the current research. To mimic the tidal movement Reynolds required a motor and pointed briefly to the efforts required to keep his model running:

"At the highest speed, 2 tides a minute, the motor only makes about 200 revolutions per minute, so that the 13,000 gallons will keep it going over three days, and has done so from Saturday till Tuesday, Monday being Bank Holiday" Reynolds (1887).

More than hundred years of scientific progress shows that the problem of estuarine morphodynamics remains intriguing by its complexity. Models consisting of real water and sand have been replaced by computers, but we still need continuous attention and patience to keep these models running (sometimes over several months). And not only on Bank Holiday.

Perhaps I should start by acknowledging the people that provided most important support over the past years when I was running these models and presenting their outcomes all over the world during an ever busy family life. Anneli has been a great support and continuous inspiration allowing for countless out-of-office hours that I spent on this PhD study. I pay tribute to our kids, Zide, Ben and Inez, who had to deal with a father who was often absent-minded and regularly traveling abroad. Thanks to my parents (Gerty and Rob van der Wegen) for their trust and encouragement. Together with parents in law (Ad en Ineke Kramer) and other family members they have been *always* available for extra logistical support.

Dano Roelvink has proven to be a 'perfect match' in the supervision of this thesis. Apart from his known expertise on the modeling aspects, his personal guidance more than once 'exactly hit the button'. Thank you, room-mates Dano Roelvink and Ali Dastgheib, who made every day working life enjoyable and allowed me space and time to develop and elaborate new ideas. Thank you, other colleagues, who supported me with small but highly valued comments. Other people that I want to mention here contributed to the work by inspiring feedback or technical support: Bruce Jaffe, Dave Schoellhamer, Neil Ganju (USGS), Ian Townend (HR Wallingford), Huub Savenije, Henk Jan Verhagen (TUD), Edwin Elias, John de Ronde, Zheng Bing Wang, Maarten van Ormondt, Arjen Luijendijk, Bert Jagers (Deltares) and others whom I unintentionally forget. I thank all students that I could guide during their MSc and PhD study over the past years at UNESCO-IHE and who indirectly inspired me to start and complete this study.

I acknowledge USGS Priority Ecosystem Studies and CALFED for financial support for the work described Chapter 6 carried out within the framework of the CASCaDE climate change project (CASCaDE contribution 17). Finally, I acknowledge UNESCO-IHE for the commitment to this work and the UNESCO-IHE research fund for making this research financially possible.

Mick van der Wegen

Table of Contents

1 Framework for morphodynamic modeling in alluvial estuaries

Abstract

The following chapter elaborates on the research framework of this study. It starts with an overview of the (historically) observed morphodynamic behavior in estuaries. This ranges from phenomenological observations to empirical equilibrium relationships and the concept of an 'ideal' estuary. These observations emphasize the importance of improving process knowledge on morphodynamics in estuarine systems. Within that framework, the need, possibilities and importance of morphodynamic modeling techniques are addressed.

The second part of this chapter defines the aim and main objectives of the study. Specific research questions are formulated and the relevance of the study is highlighted.

The final part introduces the Delft3D modeling system, which is the main tool in the current research. The general and governing equations and modeling techniques are specified. In principle, upcoming chapters apply these model settings and specify occasional adaptations or additional techniques.

1.1 Introduction

1.1.1   *What is estuarine morphodynamics?*

Estuarine morphodynamics describes the development of the estuarine bed over time. The tidal movement and waves stir up sediment from the bed and deposits the material again in more sheltered areas at times during the tidal cycle in which there is hardly any water movement. The continuous and long-term interaction between the tide, waves and sediments shapes the bed and results in an estuarine morphology, i.e. the characteristic morphological patterns often found in estuarine systems like ripples, dunes, shoals and delta's [Dalrymple and Rhodes (1995), Hibma et al. 2004)]. As an example, Figure 1.1 shows the channel and shoal pattern of the Western Scheldt Estuary.

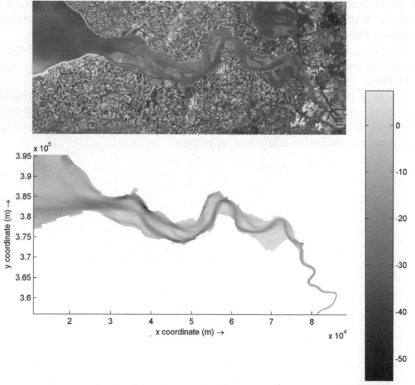

Figure 1.1 *Western Scheldt estuary. The upper panel shows a satellite image of the geometry as well as other topographic features such as urbanization and the ports of Vlissingen (upper bank towards the estuary mouth) and Antwerp (lower right corner). Shoals between the channels are clearly visible. The lower panel shows the measured 2005 bathymetry with the channels in darker colors.*

Although one may hardly notice any developments of the channel-shoal pattern in a time frame of several days, the estuarine system is not in equilibrium on the longer term. Small bathymetric developments during a tidal cycle may eventually lead to larger changes so that channels migrate or that tidal embayments are slowly silting up. Furthermore, forcing conditions at the sea or river side may not be constant on the long term. Sea level is rising and river discharge regimes may change because of climate change. The changing morphology over time is referred to as the morphodynamics of an estuarine system and is the main subject of research of this thesis.

The morphodynamics of estuaries is characterized by a complex and often nonlinear behavior. Partly this can be understood from the related hydrodynamic processes, such as turbulence generated by bed forms or the non-linear behavior of the tidal propagation in an estuary. As a result of sediment transport and changing bed levels, the development of the bathymetry itself will feed back on the hydrodynamic process. An example of this is a meandering channel that directs the flow pattern.

Characteristic morphological time scales range from hours to several decades or even hundreds of years. Typical length scales may vary from ripples and dunes to sandbanks and complete estuaries [De Vriend (1996)]. The different scale characteristics should not be seen as very distinct, since they partly overlap characteristics of other scales and contribute to continuous and sometimes cyclic processes. The complexity of the system is also illustrated by the fact that small-scale developments may have significant impact on larger scale, for example when small channels originating from dune migration develop into large channels.

1.1.2   Relevance and research scope

While carrying sediments, nutrients and pollutants, rivers flow downhill and meet the sea in an area referred to as an estuary. Estuaries are valuable areas of both local and international importance. From an ecological perspective breeding fish and migrating birds are only examples of the estuarine ecosystem richness. Furthermore, fresh and salt-water ecosystems co-exist, interact with each other and form an environment that is tidally renewed with water and nutrients. The morphology of estuaries is closely linked to these estuarine values. Tidal flats and salt marshes are essential in providing food for migrating birds and stimulate local flora and fauna.

At the same time, and because of the rich ecosystem and strategic location, estuaries have been subject to human settlement. The channel system provides a natural access to ports, although nowadays in many cases regular dredging is required to allow a sustainable access for sea vessels with increasing draughts. Numerous ports situated along estuaries form the logistical link between ocean transport and the hinterland. Further, local fishery, aquaculture and tourism are important sectors that profit from the estuarine environment.

As a result, in many estuaries there is a strong anthropogenic pressure. Visual examples are the construction of groins and dredging of access channels to ports. The morphodynamic system and the related ecosystems are artificially disturbed. Further, and on the longer term, sea level rise influences long-term behavior of the morphodynamic system and its impact on the related ecology. It is thus of major importance to develop insight into the morphodynamic processes in an estuary and their timescales in order to estimate impacts of, for example, human interference (i.e. land reclamation and dredging), sea level rise or land subsidence on the estuarine system.

First observations and research of estuarine morphology must have been done since human beings settled along the estuarine embankments and sailed with their ships for fishery and trade. Based on a review of existing literature, the following merely (but not necessarily) chronological classification is suggested considering research efforts in estuarine morphodynamics.

- Phenomenology. This earliest type of research focuses on describing observations of characteristics and developments and explaining these observations by associative reasoning;
- Empirical and equilibrium relationships. This type of research aims to empirically derive generic relationships that are valid for estuarine geomorphic features. Examples are the relation between tidal prism and cross-sectional area or channel volume in a basin or the concept of an 'ideal' estuary.
- Modeling efforts. This type of research aims to explain and distinguish in a more fundamental way the dominant processes governing estuarine morphodynamics.

These research types are discussed in the coming sections.

1.1.3   Observations and empirical research

1.1.3.1   Phenomenology

Although numerical process-based morphodynamic models were not available before the 80's of last century, several authors reported on the behavior of complete estuarine systems based on observations and phenomenological descriptions. It is remarkable how accurate these descriptions appear to be, especially in the light of current advanced, process-based numerical modeling. Examples given here focus on the channel system as well as the general geometry of alluvial estuaries.

Van Veen (1936, 1950, in Dutch) and (2002, in English) describes the different systems of ebb and flood channels based on observations along the Dutch coast. Flood channels are characterized by a bar developing at the end of the channel, whereas ebb channels deposit sediments in the ebb direction. Special interest is given to the "evasive" character of the system where both channels seem to evade one another frontally or in the flank. An idealized overview is presented (Figure 1.2 and compare to Figure 1.1) for the Western Scheldt were the sine shaped main (ebb) channel meanders through the estuary, whereas flood channels develop in the bends trying to force a more straightforward flow by cutting off the bends. Van Veen recognizes that the bends are rarely cut of and when this occurs it is of short duration and incomplete. Connections between ebb and flood channels are often classified as migrating cross channels. Van Veen states that the sill upstream of a flood channel maintains its elevation, although large water level gradients and associated large currents occur at the top of the sill. Large sand transports on a sill are the probable cause.

Another example of observation based research is the work done by Ahnert (1960) who describes the assumed origin of estuarine meanders in Chesapeake Bay, USA, bounded by Pleistocene terraces. In between the terraces the estuary meanders and an alternating bar pattern is present that consists of tidal marshes. Ahnert distinguishes ebb and flood channels in the estuarine meanders that are separated by underwater sand bars. He gives a plausible explanation for the observed area where the meandering system is located (namely in between the river head with typical fluvial marshes and the estuarine mouth with its pronounced funnel shape). Both the availability of sediments supplied by the river and the tidal distortion play a major role. He concludes that standing wave conditions, with the maximum ebb and flood velocities taking place both at mean water level, are an essential condition for the presence of estuarine meanders.

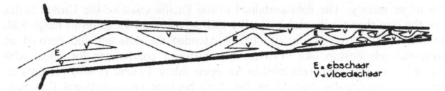

Figure 1.2 *Idealized overview of ebb and flood channel system in the Western Scheldt by Van Veen (1950).*

With respect to the general geometry of alluvial estuaries and based on empirical research over the world Savenije (1992a) suggests that
- the width convergence of an alluvial estuary can be described by an exponential function;
- the depth is nearly constant;
- along the estuary the maximum tidal velocity at spring tide is seldom more (or much less) than 1 m/s;
- the time lag between HW and HWS is constant along an estuary
- the tidal excursion is constant along the estuary and of an order of magnitude of 10 km at spring tide for a semi diurnal tide;
- tidal damping or amplification is modest or non-existent;

Another example of observations with respect to the general estuarine geometry concerns the relative importance of waves and tidal movement. Moving from the river dominated inner area towards the estuarine mouth at the coast, the presence of waves will be more pronounced. At the mouth, wave driven sediment transport may play a dominant role in local morphodynamic processes and morphological shapes. Hayes (1975) suggests a classification of estuaries with coarse-grained sediment based on the relative impact of the tide and wave (or storm) action:
- Micro tidal estuaries (tidal range of 0-2 m) are dominated by wave action and storm depositions;
- Meso tidal estuaries (tidal range 2-4 m) are characterized by tidal delta's and tidally induced sand bodies;
- Macro tidal estuaries (tidal range > 4m) are usually funnel shaped and wide mouthed containing linear sand bodies.

1.1.3.2 Empirical equilibrium relationships

An important and well explored area in estuarine morphological research focuses on inlet stability. The main reason is probably that access channels to ports are often situated in tidal inlets, indicating their direct relevance. As the name already suggests (inlet 'stability') the research looks for stable or equilibrium relationships that would have a general value and could be applied in designing port access channels and increase general understanding of morphodynamic estuarine characteristics.

Based on the work of Brown (1928), Escoffier (1940) derived a relationship between the cross sectional area at the inlet mouth and maximum velocities occurring in the inlet, given a certain basin length, depth and area. In relation to critical velocities, this led to a description for conditions for stable tidal inlets. Based on observations in tidal basins, O'Brien (1931, 1969) developed a relationship between the cross-sectional area at the estuarine mouth at low water and the tidal prism of an estuary. The data pertained to the Pacific coast of the United States where the tide shows a diurnal inequality and only a small variation of range with location. As a follow up, Jarrett (1976) extended the data set and derived an exponential relationship like O'Brien with different coefficients for inlets with jetties inlets and a linear relationship for open inlets. Eysink (1990) presents an overview of empirically derived relationships between cross-sectional flow area between MSL and the tidal volume, and channel volume below MSL and the tidal volume. The relationship holds both for tidal inlets in relation to the tidal prism of the entire basin and individual channels in relation to their part of the tidal prism. The derivation of the relationships was based on the work by O'Brien (1969) and observations at the Wadden Sea, the Netherlands.

Based on empirical data Walton and Adams (1976) relate the tidal prism of an inlet to the amount of sand stored in the ebb delta (or: outer bar) by an exponential function. Different coefficients are proposed for the degree of exposure to offshore wave action. They conclude that the higher the offshore wave energy is, the less sand is stored in equilibrium at the ebb delta. They recommend further research on the effect of long shore energy transport on this equilibrium. Additionally, Bruun (1978) derived a relationship for the sand volume stored at the outer delta in front of tidal inlets and the mean tidal volume of the inlet.

Marani et al (2002) present observations on the geometry of meandering tidal channels evolved within coastal wetlands characterized by different tidal, hydrodynamic, topographic, vegetation and ecological features. Although they observed a great variety of landscape carving modes features like the channel width/curvature and wavelength/width ratios prove to be remarkable constant.

1.1.4   The concept of an 'ideal' estuary

Pillsbury (1956) suggested the concept of an 'ideal' estuary which can be considered as a special equilibrium state of an estuary. In our understanding and definition, an ideal estuary describes hydrodynamic and geometric conditions that lead to a stable morphodynamic state of the estuary.

The ideal state follows from the absence of gradients in tidal water level and velocity amplitudes. This may be understood as zero gradient bed shear stresses and sediment transport amplitudes along the embayment, so that no morphological changes will occur.

In a uniform, infinitely long tidal embayment with a constant bed level, the water level amplitude decreases due to effects of friction. Landward convergence of the embayments' width, however, amplifies the water level amplitude. The counteracting effects of friction and convergence can balance each other [Jay (1991), Savenije (1992b, 1998, 2001), Savenije et al. (2008), Lanzoni and Seminara (1998, 2002) and Prandle (2003)] so that the water level amplitude remains constant and no landward gradients in sediment transport are present. Savenije and Veling (2005) show that it is not necessary that the bed level is constant, as long as the total cross-section is exponentially converging landward.

Friedrichs and Aubrey (1994), Jay (1991), Prandle (2003), and Savenije et al. (2008) analyzed the tidal wave characteristics. They observed that the tide propagates landward allowing at the same time a phase shift between maximum water levels and maximum velocities up to 90 degrees. This behavior apparently combines characteristics of both a progressive and a standing wave which is unique for strongly convergent embayments. Reflection of the tidal wave does not play a role as long as changes in cross-section are not abrupt and the solution is independent of the length of the estuary. Furthermore, these descriptions of tidal wave characteristics assume that the tidal wave amplitude is small compared to the water depth, which implies the absence of significant tidal asymmetries and only minor impact of intertidal flats, especially near the head.

An ideal estuary thus describes a landward exponentially decreasing (strongly convergent) cross-section and leads to a constant amplitude of the tidal velocity and a constant phase lag between water levels and velocities along the estuary.

Based on these assumptions, Prandle (2003) derived an expression for the length of an alluvial estuary mainly depending on the cross-sectional area at the mouth. His assumptions both include an exponentially decaying width and a linearly sloping bed level. Together with other assumptions (amongst others on the salt intrusion length) and comparing these to observations, Prandle et al. (2005) further argue that bathymetric estuarine variations are mainly caused by the forcing condition of tides and river flow and that sediment characteristics only play a minor role.

1.2 Morphodynamic modeling techniques

One may find morphological equilibrium based on the assumption of an ideal estuary and prescribing the geometry. It does not describe the time evolution of the equilibrium and neglects the impact of possible deviations from the idealized geometry and other assumptions like the negligible water level amplitude to depth ratio.

Morphodynamic modeling introduces an evolutionary time scale in the system so that developments can be visualized and understood. This can be done via physical scale models in a laboratory flume or mathematical approaches using the virtual laboratory of a computer.

1.2.1 Physical scale modeling

First documented systematic research on estuarine morphodynamics (by the author's knowledge) was carried out by O. Reynolds. In four subsequent reports [Reynolds (1887, 1889, 1890, 1891)] he published on physical scale model tests that were carried out in a laboratory flume, in order to investigate "the action of water to arrange loose granular material over which it may be flowing" reports [Reynolds (1887)]. The model set-up, configuration and research questions arising throughout the investigations show remarkable resemblance with that of the current study and the results are strikingly similar. Major findings are highlighted here.

Starting from a short, rectangular basin and a sandy flat bed [Reynolds (1889), see Figure 1.3] no final equilibrium for the longitudinal profile after approximately 17.000 tides was observed (or almost 24 days assuming a semidiurnal tide). "It thus appeared

1. That the rate of action was proportional to the number of tides;
2. That the first result of this tide-way was to arrange the sand in a continuous slope, gradually diminishing from high water to a depth about equal to the tide below low water;
3. That the second action was to groove this beach into banks and low-water channels, which attained certain general proportions.... " [Reynolds (1889)].

In a second series of tests, he introduces an estuarine geometry expanding like a V-shape towards the sea (which significantly changed the observed patterns), subsequently elongated the flume landward [Reynolds (1890), see Figure 1.4] and, finally, adding 'land-water' (discharge). In the final paper he addressed the impact of varying boundary condition with spring-neap tidal cycles (which appeared to have negligible influence) as well as the impact of a breakwater at the mouth forcing the local morphodynamics [Reynolds (1891), see Figure 1.5].

The existence of two distinct timescales (i.e. a timescale related to pattern development and a longer time scale related to the width averaged longitudinal profile) was also found by Tambroni et al. (2005) in a laboratory flume test with 'crushed hazel nuts' as a replacement for sand. Although laboratory tests seem to confirm developments similar to patterns found in nature, they still face scaling problems of ratios of friction, inertia and viscosity in the laboratory compared to reality.

1.2.2   *Framework for mathematical approaches*

Apart from physical scale models, mathematical modeling forms another approach to understand morphodynamic developments.

1.2.2.1   Reductionist and holistic approaches

Seminara and Blondeaux (2001) put forward the following perspective for (long-term) morphological modeling and distinguish between the reductionist and the holistic approach:

- The reductionist approach is based on strongly idealized input and output parameters. It originates from the idea that " ...understanding the behavior of complex systems requires that the fundamental mechanisms controlling the dynamics of its parts must be firstly at least qualitatively understood".
- The holistic approach involves detailed descriptions of the physical processes and "...assumes that the complete nature of the system can be investigated by tools that describe its overall behavior...".

The disadvantages of the reductionist and the holistic approaches seem obvious: reductionists tend to describe processes under conditions that are highly idealized and lose connection to processes in reality, whereas holists will not be able to distinguish between the impacts of different processes. However, there does not need to be a conflict between the two approaches, indeed they will strengthen each other, as long as they are truly scientific, in the sense that that any statement must be amenable to some kind of falsification procedure [Seminara and Blondeaux (2001)].

It is noted that any model can be considered as a mix of the two approaches and only a gradual distinction is made. For example, a 2D model can be based on full descriptions of the physical processes, but the fact that a 2D approach is followed instead of a 3D approach shows already that reductionist assumptions have been made.

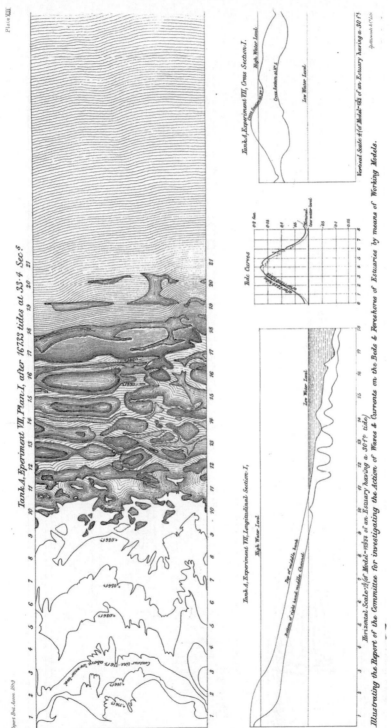

Figure 1.3 *Example of bathymetric end results of flume test with rectangular basin geometry [Reynolds (1889)].*

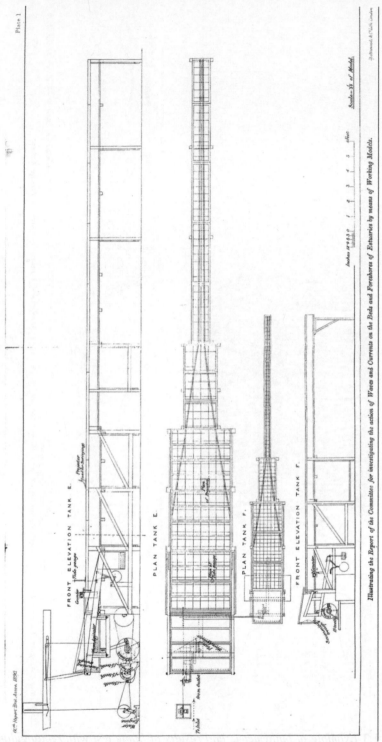

Figure 1.4 *Experimental set-up of laboratory flume with V-shaped geometry [Reynolds (1890)]*

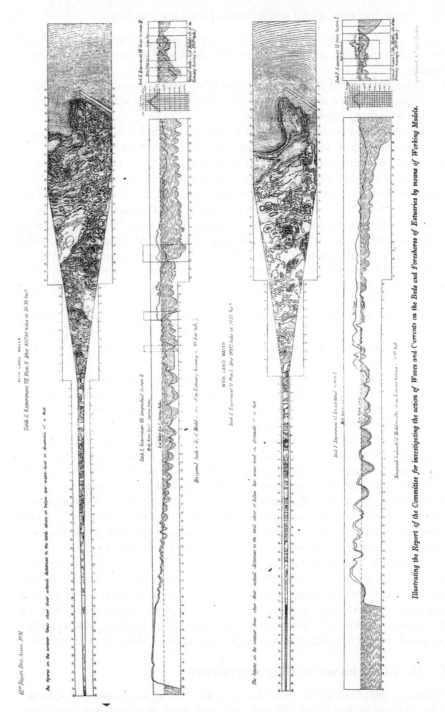

Figure 1.5 *Example of bathymetric end results of flume test with V-geometry, including effect of "land water" and inclined breakwater at the mouth [Reynolds (1891)].*

1.2.2.2   Behavior/aggregated and process-based approaches

De Vriend et al (1993b) distinguish two types of approaches to long-term mathematical morphological modeling that were developed more or less historically in sequence;

- Behavior oriented (or aggregated) modeling with a focusing on empirical relations between different types of coastal parameters without describing the underlying physical processes.
- Process-based modeling based on a detailed description of the underlying physical processes (this is the approach of the current study)

Application of a process-based model always implies that reality is reduced in such a way that no relevant processes are lost, but that, at the same time, not too many processes are included that would increase computational time too much. One may think of boundary conditions schematization (for example by applying schematized wind or wave conditions) or limiting a 3D flow to 2D or 1D modeling approaches (discussed in one of the following sections). Another way of model reduction is the application of a morphological scaling factor to speed up morphodynamic developments which are usually an order of magnitude slower than the hydrodynamic processes. This will be discussed in more detail in the following section.

A major drawback of process-based models is that (small scale) detailed description of physical processes is needed to come to (large scale) results. Especially for long-term modeling this inevitably leads to long and possibly unnecessary computation time.

Box models have been developed that start from major schematizations of the morphodynamic problem [Di Silvio (1989) and Van Dongeren and De Vriend (1994)]. Stive (1998) and Stive and Wang (2003) introduced the concept of macro scale process-aggregated modeling, embedded in the ASMITA model (Aggregated Scale Morphological Interaction between a Tidal basin and the Adjacent coast). The approach takes morphodynamic equilibrium as a starting point and assumes that a tidal basin can be schematized into a set of morphological elements (for example the ebb tidal delta, the tidal channel and the tidal flats). When all elements are in equilibrium no morphological change will take place. Disturbance of the system will lead to a morphodynamic adaptation with time scales depending on the characteristics of the individual elements and their sediment exchange relationships. These relationships are based on well-known advection diffusion equations for the sediment transport. The advantage of the approach is that it combines empirical relationships with process based sediment transport equations, which leads to time efficient modeling. The modeling approach will give insight relatively fast into the response time of macro scale impacts, such sea level rise or dredging activities, on the tidal system. Drawback of the method is that detailed morphological information and dynamics are lost and that results strongly depend on equilibrium assumptions.

1.2.3   Morphodynamic scaling techniques

Process-based morphodynamic models describe the hydrodynamics, sediment transports (including bed slope effects), and bed level developments over time. There is, however, a fundamental problem. Hydrodynamic processes take place on a time scale that is much smaller than the time scale of morphological developments. Especially for long-term morphological predictions this may lead to excessive time periods required for computation. For example, typical time steps in hydrodynamic numerical models under tidal conditions are in the order of minutes, whereas relevant morphodynamic developments only become apparent after several months. In order to make morphodynamic calculations more efficient, a number of methodologies have been developed that are described hereafter.

Using a 2D process-based model, Latteux (1995) investigated the possibility to let one tide be representative for the bed evolution of a neap spring tidal cycle or even a longer (19 years) cycle and thus reducing the input considerably. The best single tide was between mean and spring tide with peak velocities about 12 % larger than at mean tide. Additionally, Latteux investigated another way of simulating the yearly tidal cycle with any single tide by multiplying the results of different tides by a factor in such a way that the yearly evolution was reproduced properly. It was concluded that, due to the strongly non-linear relationship between flow velocity and sediment transport, bed changes mainly result from spring tides.

Additionally, there is the problem that the morphology has a much larger time scale (develops slower) than the hydrodynamic process. It would require extensive computational effort to perform a full morphological calculation after each hydrodynamic time step. Latteux (1995) also distinguishes 4 methods to reduce the number of morphological time steps;

- No flow modification as long as the bed changes do not exceed a certain threshold. This means that residual sediment transport (or the related bed level change) after one tide is multiplied by a factor N (the filtering coefficient) until the threshold value in bed change is reached.
- Like the first method only following a predictor/corrector approach where the bed change of N tides is predicted first. Based on that bed level the sediment transport and bed level change are calculated for tide N+1 using a continuity correction for the hydrodynamics. Finally, the average sediment transport and bed level change (from bed level N and N+1) are calculated resulting in a corrected bed level at N+1.
- Simulating N tides by a singly one extending N times. This method allows bed forms to propagate, but the effect of subsequent tides (chronology) is not ensured.
- Linearization of the sediment transport by expansion of the sediment transport as a function of bed evolution. Use is made by the continuity correction concept; thus, the sediment transport can be calculated for different bed level changes, without solving the full hydrodynamic equations. The method implicates that bed changes over the tide N+1 depend on bed changes of the preceding tide and on terms of the first tide.

Roelvink (2006) also compares different methodologies for morphodynamic modeling. He distinguished two main approaches, namely the Rapid Assessment of Morphology (RAM) and the Online method. RAM uses tide averaged residual sediment transports, in combination with a continuity correction and a morphological factor for updating the bathymetry. The Online method updates the bed with a morphological factor every hydraulic time step (~minutes). Comparing the different methods leads to the conclusion that Online has less restriction on the numerical stability and is more accurate than RAM applying the same morphological factor. A parallel version of Online, that allows for different parallel processes to be carried out at the same time and merging these processes for a weighted morphodynamic update will even increase the numerical stability.

Examples of case studies applying process-based models for decadal morphodynamic predictions are given by Cayocca (2001), Gelfenbaum et al. (2003) Lesser (2009) and Ganju et al. (submitted).

1.2.4   Dimensions in modeling

1.2.4.1   1D Profile modeling

An estuary can be schematized into a 1D model based on the one-dimensional Saint Venant equations (or: depth and width integrated shallow water equations). Thus, dominant hydrodynamic characteristics can be investigated [amongst others Boon and Byrne (1981), Friedrichs and Aubrey (1988), Friedrichs et al. (1990), Toffolon (2006), Savenije et al. (2008)]. Also, morphodynamics may be included to study width-averaged longitudinal profile development and equilibrium conditions. An example is the work by Schuttelaars and De Swart (1996, 2000) and Hibma (2003a) who developed an idealized 1D model of a rectangular tidal embayment to study the existence of profile equilibrium and its sensitivity to basin length, the presence of overtides and the relative strength of diffusive and advective transports. Other examples are the work by Lanzoni and Seminara (2002) who investigated the 1D morphodynamic equilibrium of a funnel shaped embayment. Drawback is that, by definition, 1D models do not resolve and account for 2D patterns (unless in a highly schematized way).

1.2.4.2   2D Modeling of channel/shoal patterns

Seminara and Tubino (2001) investigated the governing processes in the formation of sand bars in a highly schematized tidal environment. Schuttelaars and De Swart (1999), Schramkovski et al (2002) included wider basin model configurations and flow conditions that are not frictionally dominated. They demonstrated that 3d model formulation and quadratic bottom shear stress are not crucial for the initial formation mechanism. Although the model focuses on initial growth of forced perturbations, results suggest the existence of equilibrium morphological wavelengths. Schramkovski et al (2004) extended the model results to the non-linear domain so that the finite amplitude behavior of the bars could be investigated.

In a predefined and pronounced channel-shoal pattern environment (highly non-linear domain), Coevelt et al. (2003) distinguish between primary horizontal residual flow (in longitudinal direction) and secondary flow in transverse direction (usually known as spiral flow and defined as flow normal to depth averaged flow). Results show that the water levels are relatively high at the outer bend and relatively low on the shoals and that the effect of primary horizontal residual flow is an order of magnitude higher than the secondary transverse residual flow. It was concluded that depth averaged (2D) flow formulation would suffice for the morphological prediction, at least, in the Strait basin modeled.

Hibma et al. (2003b,c) and Marciano et al (2005) shows the development and evolution of a channel-shoal system in elongated (Hibma) and back barrier (Marciano) tidal embayment and by means of a process-based, numerical model (Delft3D) and starting from a flat bed. Long-term simulations (carried out up to 300 years) enter the non-linear domain were comparison to models based on initial channel shoal development [Seminara and Tubino (2001), Schuttelaars and de Swart

(1999), Schramkovski (2002)] would not be possible anymore. The initial development of perturbations is characterized by relatively small wavelengths, which finally develop to wavelengths with an order of magnitude of the tidal excursion length. Hibma et al. (2003b, c) report that dominant morphological wavelengths decrease further landward, because of the decreasing tidal prism. The model results suggest that the channel-shoal patterns evolve towards equilibrium in a century time scale with dominant wavelengths independent of the initial perturbations. The applied sediment transport formula and grain size mainly influence the time scale but not the large-scale pattern development. Furthermore, quantitative comparison of the number of channels and meander length scales with observations (Western Scheldt, the Netherlands, and Patuxent River estuary, Virginia) give good accordance. Dominant wavelengths are found to be somewhat larger (10%) than the initial dominant wavelength. This is explained by the fact that the maximum velocities in the channels increase and the effect of friction is less, as shoals and channels develop (this results in a larger tidal excursion in the deeper channels). Other non-linear effects are probably, amongst others, the effect of the shoal channel system directing the main flow, siltation of the embayment reducing the tidal prism, and a changing distortion of the tide.

1.3 Aim of the study

1.3.1 Main objective

The main objective of this research is to investigate the governing processes and characteristics that drive morphodynamic evolution in alluvial estuaries by application of a process-based numerical model.

1.3.2 Research questions

Taking the literature survey as a starting point, 5 research questions are articulated to further specify the main research objective:

1. How can morphodynamic 'equilibrium' be defined?
Morphodynamic 'equilibrium' is a questionable concept. Although equilibrium has been reported in modeling of highly schematized environments and in empirical relationships it is not clear whether or not equilibrium exists in reality and on longer time scales. It is therefore useful to address the evolutionary time scale and indicators for equilibrium.

2. What timescales can be distinguished in long-term, 2D process-based morphodynamic modeling?
Under highly idealized conditions 1D models show the development of equilibrium longitudinal (or: width averaged) profiles. 2D modeling of channel-shoal patterns shows growth rates of preferred morphological wavelength in the linear domain and even equilibrium patterns in the non-linear domain. No detailed study was done on how these two morphologic characteristics interact and whether or not they have separated time scales.

3. What processes are relevant in long-term morphodynamic evolution?
Literature describes numerous detailed processes (i.e. flocculation, turbulence, current wave interaction, biological impact on critical shear stresses) which are not fully understood and known in a model domain. One may argue that only inclusion of these processes may lead to accurate morphodynamic predictions. This would lead to excessive computation time. On the other hand, alluvial estuarine systems may be governed by dominant forcing mechanisms which allow for reduction of processes and input. The question is to what extent the model may be reduced and what processes are of second order in describing morphodynamic evolution.

4. Does long-term morphodynamic modeling under free and idealized conditions lead to a preferred geometry?
The concept of an 'ideal' estuary assumes certain geometric qualities of the estuary based on observations. Examples are a constant bed level or a landward exponentially decaying cross-sectional area and related constant water level amplitude and velocity amplitude along the embayment. However, no investigations have been reported that describe the morphodynamic evolution towards such geometry and conditions.

5. What is the influence of geometry on the morphodynamic characteristics of an estuary?
In studying evolutionary morphodynamic timescales and characteristics it is useful to apply idealized circumstances and geometries. However, these conditions may differ considerably from those found in reality. Apart from more complex boundary conditions, the impact of the geometry on morphodynamic development might be significant. It is worthwhile to investigate the role of geometry compared to the impact of model definitions like sediment characteristics and transport formulations.

1.3.3 Relevance

Originally, the objective of this study was triggered by the non-availability of bathymetric data in developing countries. It was felt that there was a need for models that could predict the bathymetry to such an extent that the model results could be used in estimating the impact of for example dredging activities or land reclamation projects in an estuary.

An example may illustrate this. The access channel to a port in a remote estuary is subject to siltation. In order to predict the amount of sand to be dredged annually a numerical model can be made. However, no trustworthy morphological data are available. Still a model can be developed having only few data available. Predictions on tidal water levels can be derived from easily available models that are based on harmonic analysis of the tidal constituents. Also, river discharges are usually available via measurement stations. Roughly, the shape of the estuary can be known from remote sensing (satellite) data. Based on these boundary conditions a numerical model could be developed that describes the hydrodynamics, the sediment transport and the resulting bathymetry from an initial flat bed. The question is then to what extent the modeled bathymetry would describe the real bathymetry and how the difference between model bed and reality could be minimized.

By developing a morphodynamic model for the longer term, inevitably system knowledge will be gained on morphodynamic equilibrium present in estuaries. The research will thus contribute to the long-term behavior of all estuarine systems, situated in developing countries or not.

1.3.4 Approach

1.3.4.1 Modeling tool

The current study applies a process-based model (Delft3D). The model numerically solves the shallow water equations, so that water levels and velocities are described in the modeled area at any time and at any place in the grid. These data are used to predict the associated sediment transport. The change of the bathymetry is subsequently calculated from the divergence of the sediment transport field in the area. Thus, the morphodynamic development originates from evolving interactions between flow and bed level development. Flow characteristics shape the bed and bed level developments, in their turn, direct the flow. The details of the model are described in the following section.

1.3.4.2 Strategy

The thesis starts from a relatively simple model set up with highly schematized model configuration (a rectangular tidal basin filled with uniform sand and forced by a semidiurnal tide) to study characteristic morphodynamic evolution over millennia. In subsequent chapters more processes are added, free behavior is restricted and the timescale is reduced, to finally end up with a model configuration that matches the geometry of a real alluvial estuary and a 'comprehensive' timescale that allows model validation with measured morphological developments. Figure 1.6 shows the schematized thesis outline.

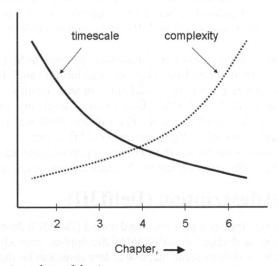

Figure 1.6 *Schematic outline of thesis.*

In 1889 Reynolds argued the reason why he choose his approach in research, which appears strikingly similar to the strategy of the current research "... it appeared that as the main object of these researches is to differentiate and examine the various circumstances which influence the distribution of the sand, it was desirable, in starting, to simplify as much as possible all the circumstances directly under control, and so afford an opportunity for other more occult causes to reveal themselves through their effects" Reynolds (1889).

1.3.4.3 Chapter outlines

Chapter 1 includes the research framework, literature review, and the aim of this study and a description of the applied process-based model (Delft3D).

Chapter 2 investigates 3200 years of morphodynamic evolution in an 80 km long and 2.5 km wide rectangular tidal embayment under constant semidiurnal forcing conditions. 2D Model results are compared to 1D results and sensitivity analysis is carried out by varying the width and length of the basin as well as the value of the morphological factor.

Chapter 3 adds a bank erosion mechanism to the model configuration of chapter 1 so that both bathymetry and geometry of the basin can develop. The timescale is about 6400 years. Results are evaluated in terms of energy dissipation.

Because of the long timescales involved and the highly schematized model set-up validation of the model results of Chapter 2 and 3 is difficult. One of the possibilities of model validation is by comparing results to empirical equilibrium relationships that seem to have general value. Therefore Chapter 4 focuses on comparison of the model results of Chapter 2 and 3 with empirically derived relationships between the tidal prism and the cross-sectional area.

Chapter 5 investigates the role of the geometry in the development of morphological characteristics. Focus is on the Western Scheldt estuary. Central question in this chapter is to what extent Delft3D is able to reproduce the 1998 channel shoal patterns starting from a flat bed (which includes the same amount of sediment as the 1998 measured profile) and restricting the model domain by the 1998 Western Scheldt geometry. Sensitivity analysis is carried out on the presence of different tidal constituents, 2D and 3D grids, non-erodible layers and dredging and dumping scenarios.

Chapter 6 applies the process-based modeling approach to San Pablo Bay (which is a sub-bay of San Francisco Bay in California) and hindcasts decadal morphodynamic development by validating model results against measured bathymetric data in 1856 and 1887. Due to the complexity of the area proper hindcast requires inclusion of processes like salt and fresh water interaction, wind, waves and the application of multiple sand and mud fractions.

Chapter 7, finally, summarizes the findings of previous chapters and relates these to the original aim of the research and the 5 detailed research questions.

1.4 Model description (Delft3D)

The current research applies a process-based model (Delft3D) described by Lesser et al (2004) in which a detailed description of the applied hydrodynamic equations, numerical aspects, as well as some elaborated test cases can be found. However, the focus in Lesser et al (2004) is on a description of the 3D mode, although it also includes a comparison of the model run in 2D (depth-averaged) and 3D mode for some case studies. Since the current research merely focuses on a 2D depth-averaged situation, in the following sections the main model characteristics will be described. This concerns, firstly, the governing hydrodynamic equations and, secondly, the formulations describing the morphodynamics. Details are presented on numerical aspects of the model, in particular with respect to the applied morphodynamic update scheme, dry cell (or: bank-) erosion and a specification of flooding and drying criteria. Finally, a small introduction is presented on the Brier Skill Score (BSS). This parameter is a useful tool to assess the performance of morphodynamic models. It is applied in the final two chapters describing modeling results of case studies of the Western Scheldt, the Netherlands, and San Pablo Bay, California.

The chapters of this thesis are based on the following general model formulations. Deviations (for example in case of a different sediment transport formulation) will be addressed in the specific chapters.

1.4.1 Hydrodynamic model

The hydrodynamic model is unsteady and two-dimensional and is based on the set of shallow water equations, in which vertical velocities are neglected. Neglecting the influence of the Coriolis' force, density differences, wind and waves, the two dimensional equation of continuity and the momentum equations read as follows:

$$\frac{\partial \zeta}{\partial t} + \frac{\partial h\bar{u}}{\partial x} + \frac{\partial h\bar{v}}{\partial y} = 0 \tag{1.1}$$

$$\frac{\partial \bar{u}}{\partial t} + \bar{u}\frac{\partial \bar{u}}{\partial x} + \bar{v}\frac{\partial \bar{u}}{\partial y} + g\frac{\partial \zeta}{\partial x} + c_f \frac{\bar{u}\sqrt{\bar{u}^2 + \bar{v}^2}}{h} - v_e\left(\frac{\partial^2 \bar{u}}{\partial x^2} + \frac{\partial^2 \bar{u}}{\partial y^2}\right) = 0 \tag{1.2}$$

$$\frac{\partial \bar{v}}{\partial t} + \bar{v}\frac{\partial \bar{v}}{\partial y} + \bar{u}\frac{\partial \bar{v}}{\partial x} + g\frac{\partial \zeta}{\partial y} + c_f \frac{\bar{v}\sqrt{\bar{u}^2 + \bar{v}^2}}{h} - v_e\left(\frac{\partial^2 \bar{v}}{\partial x^2} + \frac{\partial^2 \bar{v}}{\partial y^2}\right) = 0 \tag{1.3}$$

With

$$c_f = g\frac{n^2}{\sqrt[3]{h}} \tag{1.4}$$

In which

ζ water level with respect to datum, m
h   water depth, m
$\bar{u}$ depth averaged velocity in x direction, m/s
$\bar{v}$ depth averaged velocity in y direction, m/s
g   gravitational acceleration, m²/s
c_f friction coefficient, -
n   Manning's coefficient, $sm^{-1/3}$
v_e eddy viscosity, m²/s

If not otherwise stated in this research the applied value of Manning's coefficient is 0.026 $sm^{-1/3}$ and the value of the eddy viscosity, v_e, is 0.1 m²/s.

1.4.2 Sediment transport formulation

The velocity field obtained by solving the equation of continuity and the momentum equations is used to calculate the sediment transport field. Various sediment transport formulations are available for this in literature. Van Leeuwen and De Swart (2004) and Schuttelaars and De Swart (1996, 2000) indicated the different character of a 1D equilibrium profile for different values of diffusive and/or advective transports. Further, based on a literature review, Lanzoni and Seminara (2002) extensively analyze and describe dominant sediment transport processes in tidal environments. These are (amongst others) the presence of multiple sediment fractions including sand and mud, tidal asymmetries in currents and water levels resulting in continuously changing suspended sediment concentration profiles during the tidal cycle and erosion and settling lags particularly relevant at shoals around slack tide. Despite the recognition of these processes, for their model Lanzoni and Seminara (2002) finally choose for formulations of the bed and suspended load transports based on local and instantaneous flow conditions, because their analysis showed that these formulations would be justified at least in terms of leading order effects.

The current research follows this reasoning and the authors, additionally, choose for a relatively simple transport formulation without making distinction between suspended load and bed load. Thus, results could be analyzed in a more straightforward way whereas future research should focus on more complex transport formulations.

Different authors (amongst others Groen (1967), Lanzoni and Seminara (2002), Prandle (2004), Van Leeuwen and De Swart (2004) and Pritchard (2005)) have pointed to the importance of suspended sediment transport for prediction of equilibrium conditions in a tidal embayment, especially regarding the settling lag effect. However, it is argued that processes of bringing sediment into suspension and settlement of suspended sediments take place within the grid cells of the current model (about 100 by 100 m) and do not significantly affect the amount of transported sediments. Assuming that advection dominates diffusive transports and a water depth (h) of 8m, a characteristic flow velocity (u) of 1 m/s and a fall velocity (w) of 0.02 m/s, the length (dx) for a sediment particle at the surface to settle at the bed and being advected by the flow at the same time is dx=hu/w=400m. This value is an upper value considering that suspended sediment is mainly concentrated in the lower part of the water column, that the water depth is lower than 8m in large parts of the current basin and that velocities are typically lower than 1 m/s during the tidal cycle.

Pritchard (2005) confirms that for relatively large ratios of the tidal time scale to the time scale over which the sediment concentrations respond to changes in erosion and deposition rates, lag effects are small. This holds for the current research with a uniform sediment size of 240 μm. Preliminary calculations with transport formulations including a distinction between suspended load and bed load showed results that are comparable to the results presented in the current research. Finer sediments would increase the settling lag effect and probably induce a nett landward transport. These assumptions are supported by the observation that preliminary calculations including distinct formulations for the bed load and suspended load (i.e. Van Rijn (1993)) indicate no large differences from results based on the current Engelund-Hansen formulation.

Use is made of the instantaneous total sediment transport formula developed by Engelund-Hansen (1967) that relates velocity directly and locally to a sediment transport:

$$S = S_b + S_s = \frac{0.05U^5}{\sqrt{g}C^3\Delta^2 D_{50}} \tag{1.5}$$

where

S	magnitude of the sediment transport, m^3/ms
S_b, S_s	magnitude of the bed load and suspended load transport, m^3/ms
U	magnitude of flow velocity, m/s
Δ	relative density $(\rho_s - \rho_w)/\rho_w$, -
C	friction parameter defined by $\dfrac{\sqrt[6]{h}}{n}$, m$^{1/2}$/s
n	Manning's coefficient, sm$^{-1/3}$
D_{50}	median grain size, m.

1.4.3   Morphodynamic model

1.4.3.1   Bed slope effect

Usually, distinction is made between bed load and suspended load, where the effect of the bed slope on the bed load transport is at least an order of magnitude larger than the bed slope effect on the suspended load transport. However, the current model does not distinguish between bed load and suspended load since a total sediment transport formula is used. Also, the current model uses a formulation for the bed slope effect that was developed for bed load transport only. Thus, the total sediment transport is treated as if it were bed load. Since suspended transport forms generally the majority of the total transport in tidal embayments, the bed slope effect is thus somewhat overestimated.

The adjustment of sediment transport for bed slope effects is executed into two steps. Firstly, the *magnitude* of the sediment transport vector is adjusted in case of a bed slope in the same direction as the initial sediment transport vector, defined here as the longitudinal direction (s). It is noted that in the current model the initial sediment transport vector has the same direction as the depth averaged flow velocity vector. The size of the adjustment is calculated following a modified form of the expression suggested by Bagnold (1966):

$$\vec{S'} = \alpha_s \vec{S} \tag{1.6}$$

where

$$\alpha_s = 1 + \alpha_{bs} \left[\frac{\tan(\phi)}{\cos\left(\tan^{-1}\left(\frac{\partial z_b}{\partial s}\right)\right)\left(\tan(\phi) - \frac{\partial z_b}{\partial s}\right)} - 1 \right] \tag{1.7}$$

$\vec{S}$ initial sediment transport vector, $m^3/m/s$

$\vec{S'}$ adjusted sediment transport vector, $m^3/m/s$

α_{bs} coefficient with default value of 1,-.

ϕ internal angle of friction of bed material (assumed to be 30^0), ^{0}

$\frac{\partial z_b}{\partial s}$ bed slope in longitudinal direction, -

And

$$\left(\frac{\partial z_b}{\partial s}\right)_{max} = 0.9\tan^{-1}(\phi) \tag{1.8}$$

Secondly, the *direction* of the bed slope transport vector is adjusted for a bed slope in the direction normal to the initial sediment transport vector, referred to as (n). This is done by calculating an additional sediment transport vector perpendicular to the initial sediment transport vector. The magnitude of this vector is calculated using a formulation based on the work of Ikeda (1982) and Ikeda and Aseada (1983) as presented by Van Rijn (1993):

$$S_n = \left|\vec{S'}\right| \alpha_{bn} \frac{u_{cr}}{|\vec{u}|} \frac{\partial z_b}{\partial n} \tag{1.9}$$

where

S_n magnitude of additional sediment transport vector. The direction of this vector is down slope, $m^3/m/s$

α_{bn} coefficient, default value = 1.5, -

u_{cr} critical (threshold) depth-averaged flow velocity, m/s

$\vec{u}$ depth-averaged flow velocity vector, m/s

$\frac{\partial z_b}{\partial n}$ bed slope in the direction normal to the initial sediment transport vector, -

The resulting sediment transport vector ($\vec{S_r}$), which is adjusted for magnitude and direction) is calculated by:

$$\vec{S_r} = \vec{S'} + \vec{S_n}$$ (1.10)

1.4.3.2 Bed level update

Conservation of sediment is described by the following equation representing a balance between the divergence of the sediment transport field and the evolution of the bed level corrected for bed porosity:

$$\left(1-\varepsilon\right)\frac{\partial z_b}{\partial t} + \frac{\partial S_x}{\partial x} + \frac{\partial S_y}{\partial y} = 0$$ (1.11)

where

ε bed porosity, default 0.4, -
z_b bed level, m
S_x sediment transport in x-direction, $m^3/m/s$
S_y sediment transport in y-direction, $m^3/m/s$

Following equation (1.11) the bed level in the model is updated every time step based on a constant porosity and sediment transport gradients along a grid cell. Next section elaborates on a technique to enhance morphodynamic developments.

1.4.3.3 Morphodynamic update

Tidal hydrodynamic behaviour in estuaries has a time scale that is typically 1 to 2 orders of magnitude smaller than the morphodynamic time scale (Wang et al. (1991)). In terms of numerical modelling this implies that many hydrodynamic calculations need to be performed that have only limited effect on the morphology. It is only after several weeks that the impact on the bed becomes relevant. Morphodynamic calculations would therefore require long and inefficient hydrodynamic calculation time. In order to increase the efficiency of process-based morphodynamic models different techniques have been developed.

Latteux (1995) describes, amongst others, the concept of the morphological tide that uses the residual sediment transport calculated over one tide to update the bed for a next hydrodynamic calculation. Roelvink (2006) compares a tide averaging method with continuity correction (RAM) to the "online" method and concludes that the "online" approach is most favorable. In this approach the bed level change calculated every hydrodynamic time step is increased by a morphological factor and results in the bed level used in the next hydrodynamic time step. Model results are valid as long as bed level changes after a time step remain small compared to the water depth. The update scheme is presented in Figure 1.7. Main advantage is that the stability and accuracy of the method are less restrictive than for the tidally averaged method (RAM) and that it allows propagation of morphological features during a tide.

In our model the "online" approach is applied with a morphological factor (MF) of 400. The sensitivity tests carried out by Roelvink (2006) are based on tidal inlet geometry somewhat differing from the current research geometry. It is, however, expected that the allowed value of the morphological factor will depend on the modeled processes (i.e. waves and/or tides) and model configuration. Its final value needs to be determined per case. The results of an extensive sensitivity analysis of different values of the MF are discussed more closely in upcoming sections.

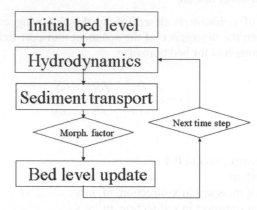

Figure 1.7 *Morphological update scheme*

1.4.3.4   Dry cell erosion

The advantage of using a morphological factor is that it accelerates the morphodynamic calculations in such a way that the computational effort can be drastically decreased. The disadvantage of high values of the morphological factor is that it may lead to a wrong description of the physical processes, because erosion and deposition processes are extrapolated. It increases the risk of cells becoming permanently 'dry'. At certain locations and instances the water depth remains below the threshold value for wet cells described in the previous section or in a bed level, which exceeds high water. These cells are defined to become dry. The cells with bed levels exceeding high water even develop into fixed points that cannot participate in the erosion processes anymore and undermine the morphodynamic character. The problem is not only apparent for the 'online' approach, but holds for all methods that apply an accelerated bed level update. For example, Hibma (2003b) experienced the development of fixed points at the shoals using a similar configuration as the current research only with bed level updates based on tidally averaged transports.

The problem is overcome by allowing erosion of dry cells in such a way that the erosion that is taking place in a wet cell is assigned to the adjacent dry cell. Implicitly this means that sediment is transported from a dry cell to a wet cell so that no bed level change of the wet cell takes place. The procedure continues until the dry cell becomes wet again. The procedure is not solely applied to cells with a bed level exceeding the highest water level during the tide but also for cells becoming dry during the tidal cycle. The procedure is more extensively described and applied in Roelvink et al (2003).

The following example of intertidal area becoming dry during the tide may illustrate the combined processes of drying and flooding and dry cell erosion. Assume a falling tide which results in a particular cell becoming dry when the water depth becomes lower than 0.1m. No erosion of the cell can take place anymore due to transport processes within the cell. However, erosion may take place when the adjacent wet cell erodes. This erosion is attributed to the dry cell, which may become wet again when its water depth exceeds 2*0.1m. The erosion of the dry cell should then be considered as resulting from bank erosion effects at the boundary between the wet and the dry cell rather than from erosion by hydrodynamic processes within the dry cell.

1.4.4 *Numerical aspects*

1.4.4.1 Hydrodynamics

The numerical scheme applies an orthogonal, staggered grid, where water level points and depth-points are co-located in the cell centers and the u- and v- velocity points are located in the middle of the cell walls. An Alternating Direction Implicit (ADI) method is used to solve the continuity and momentum equations (Leendertse, 1987). The advantage of the ADI method is that the implicitly integrated water levels and velocities are coupled along grid lines leading to systems of equations with a small bandwidth. Stelling (1984) extended the ADI method of Leendertse with a special approach for the horizontal advection terms, namely the splitting of a third order upwind finite difference scheme for the first derivative into two second order consistent discretisations: a central discretisation and an upwind discretisation, which are successfully used in both stages of the ADI-scheme. The scheme is denoted as the "cyclic method" (Stelling and Leendertse, 1991).

1.4.4.2 Morphodynamics

General
With respect to the bed slope effects on the sediment transport described in the previous section, the longitudinal and normal bed slopes are calculated by

$$\frac{\partial z_b}{\partial s} = \frac{\partial z_{(u)}}{\partial x}\frac{S_x}{\left|\vec{S}\right|} + \frac{\partial z_{(v)}}{\partial y}\frac{S_y}{\left|\vec{S}\right|} \tag{1.12}$$

$$\frac{\partial z_b}{\partial n} = \frac{\partial z_{(u)}}{\partial x}\frac{S_y}{\left|\vec{S}\right|} + \frac{\partial z_{(v)}}{\partial y}\frac{S_x}{\left|\vec{S}\right|} \tag{1.13}$$

with

$\dfrac{\partial z_{(u)}}{\partial x}$ bed slope in the positive x direction evaluated at the U point, -

$\dfrac{\partial z_{(v)}}{\partial y}$ bed slope in the positive y direction evaluated at the V point, -

$\vec{S}, S_x, S_y$ the initial sediment transport vector, the initial sediment transport in x- and y-direction respectively, $m^3/m/s$

      With respect to the morphodynamic model the following important procedures, slightly deviating from Lesser et al (2004), where followed:

- the depths in u- and v-points are taken as the *minimum* of the surrounding depths in water level points;
- the velocity vectors applied in the centers are determined by a depth-weighted averaging of the surrounding velocities in u- and v-points;
- the sediment transport components in the u- and v-points are copied from the upstream water level points where the bed load transport is evaluated;
- the bottom change (in water level points) over half a time step is computed as the net sediment transport into or out of a cell, multiplied by the morphological factor;

It is stressed that bed level changes at the mouth boundary may take place for outgoing (seaward) flow. In contrast, no bed level changes are allowed for ingoing (flood) flow at the mouth boundary cells.

Drying and flooding
The intertidal area in the model falls dry and becomes wet during the tidal cycle. Therefore a proper description of the drying and flooding procedure is needed. Bates and Hervouet (1999) provide a review of techniques on the matter and Lanzoni and Seminara (2002) used the work of Defina (2000) in their model. This approach is based on a statistical description regarding sub grid bed level variations of shallow cells including the effect of advection and friction.

The current research applies the method described by Hibma et al (2003b, c) in which the cells that fall dry are removed from the hydrodynamic calculation. When the tide rises and the cells become wet again, they are reactivated. Cells become dry when the water depth decreases below a certain threshold value (0.1 m in this study). This means that the velocity is set to zero. The cell is closed at the side normal to the velocity. When all four velocity points of a cell surrounding a water level point are dry, the cell is excluded from computation. If the water level rises and becomes larger than twice the threshold level (2*0.1m in this study), the velocity point is reactivated.

1.4.5   Brier Skill Score (BSS)

In order to assess the skill of morphodynamic models Sutherland et al (2004) suggest the use of the Brier Skill Score (BSS). For the current study the BSS is defined as follows:

$$BSS = 1 - \frac{\left\langle (\Delta vol_{\mathrm{mod}} - \Delta vol_{\mathrm{meas}})^2 \right\rangle}{\left\langle \Delta vol_{\mathrm{meas}}^2 \right\rangle} \qquad (1.14)$$

in which
Δvol volumetric change compared to the initial bed , (m^3)
mod   modeled quantity,
meas   measured quantity
and the $\left\langle \ \right\rangle$ denote an arithmetic mean. A BSS of 1 gives the perfect modeling result, whereas lower values suggest less adequate modeling. Sutherland et al (2004) note that the BSS is unbounded at the lower limit and that the BSS can be extremely sensitive to small changes when the denominator is low.

The origin of a particular BSS value may be attributed to an amplitude error, a phase error, or a deviation from the average value. In the first case an error is made in the 'height' of a particular morphological feature, in the second case the BSS is determined by a shift in location of the morphologic feature, whereas in the latter case the error originates from a deviating mean bed level. In order to assess these errors separately, Murphy and Epstein (1989) suggest decomposing the BSS as follows:

$$BSS = \frac{\alpha - \beta - \gamma + \varepsilon}{1 + \varepsilon}$$
(1.15)

in which

$$\alpha = r^2_{X'Y'}$$

$$\beta = \left(r_{X'Y'} - \frac{\sigma_{Y'}}{\sigma_{X'}} \right)^2$$

$$\gamma = \left(\frac{\langle Y' \rangle - \langle X' \rangle}{\sigma_{X'}} \right)^2$$
(1.16)

$$\varepsilon = \left(\frac{\langle X' \rangle}{\sigma_{X'}} \right)^2$$

and
r   correlation coefficient
σ   standard deviation
X' Δvol_{meas} , m^3
Y' Δvol_{mod} , m^3

Following Sutherland et al. (2004) 'α' is a measure of phase error and perfect modeling of the phase gives α = 1. 'β' is a measure of amplitude error and perfect modeling of phase and amplitude gives β = 0. 'γ' is a measure of the mean error when the predicted average bed level is different from the measured bed level and perfect modeling of the mean gives 'γ'=0. 'ε' is a normalization term, which is only affected by measured changes from the baseline prediction.

In order to account for the effect of measurement errors Van Rijn et al. (2003) suggest the following extended BSS:

$$BSS_{vR} = 1 - \frac{\left\langle \left(|\Delta vol_{mod} - \Delta vol_{meas}| - \delta \right)^2 \right\rangle}{\left\langle \Delta vol_{meas}^2 \right\rangle}$$
(1.17)

in which δ (m^3) is the volumetric measurement error and in which $\left|\Delta vol_{mod} - \Delta vol_{meas}\right| - \delta$ is set to zero if $\left|\Delta vol_{mod} - \Delta vol_{meas}\right| < \delta$. Further Van Rijn et al. (2003) proposed a classification of BSS and BSS_{VR} values as presented in Table **1-1**.

	BSS	BSS_{VR}
Excellent	0.5-1.0	0.8-1.0
Good	0.2-0.5	0.6-0.8
Reasonable	0.1-0.2	0.3-0.6
Poor	0.0-0.1	0.0-0.3
Bad	< 0.0	< 0.0

Table 1-1 *BSS classification*

2 Schematized 1D and 2D model results and Western Scheldt data[1]

Abstract

The research objective of this chapter is to investigate long-term (~millennia) evolution of estuarine morphodynamics. Use is made of a 2D, numerical, process-based model. The standard model configuration is a rectangular 80 km long and 2.5 km wide basin. Special emphasis is put on a comparison between 1D and 2D model results. Furthermore, 2D model results are compared to Western Scheldt characteristics and empirical relationships. Special attention is given to an analysis of the numerical morphodynamic update scheme applied.

Equilibrium conditions of the longitudinal profile are analyzed using the model in 1D mode. 2D model results show two distinct time scales. The first time scale is related to pattern formation taking place within the first decades and followed by minor adaptation according to the second timescale of continuous deepening of the longitudinal profile during 1600 years. The resulting longitudinal profiles of the 1D and 2D runs are similar apart from small deviations near the mouth. The 2D results correspond well to empirically derived relationships between the tidal prism and the channel cross-section and between the tidal prism and the channel volume. Also, comparison between the current model results and data from the Western Scheldt estuary (in terms of bar length, hypsometry, percentage of intertidal area and values for the ratio of shoal volume and channel volume against the ratio of tidal amplitude and water depth) shows satisfying agreement.

Based on the model results a relationship for a characteristic morphological wavelength was derived based on the tidal excursion and the basin width and an exponentially varying function was suggested for describing a dimensionless hypsometric curve for the basin.

---

[1] An edited and slightly adapted version of this chapter was published by AGU. Copyright (2008) American Geophysical Union. To view the published open abstract, go to http://dx.doi.org and enter the DOI.

Van der Wegen, M., and J. A. Roelvink (2008), Long-term morphodynamic evolution of a tidal embayment using a two-dimensional, process-based model, J. Geophys. Res., 113, C03016, doi:10.1029/2006JC003983.

2.1 Introduction

Estuaries are valuable areas of both local and international importance. Breeding fish and migrating birds are only examples of the estuarine ecosystem richness. From an economic point of view, local fishery, aquaculture and tourism are important sectors that profit from the estuarine environment. Additionally, numerous ports situated along estuaries form the logistical link between ocean transport and the hinterland.

The morphology of estuaries is closely linked to these estuarine values. Tidal flats and salt marshes are essential in providing food for migrating birds and stimulate local flora and fauna. At the same time, the channel system provides natural access to ports, although in many cases regular dredging is required to allow a sustainable access for sea vessels with increasing draughts. It is thus of major importance to develop insight into the morphodynamic processes in an estuary in order to estimate impacts of, for example, human interference (i.e. land reclamation and dredging) and sea level rise.

2.1.1   *Earlier research*

Phenomenological descriptions of estuarine morphology can be found in Van Veen (1936, 1950) and Ahnert (1960). Additionally, O'Brien (1969), Jarrett (1976) and Eysink (1990) present empirical relationships between the minimum cross-sectional area of a tidal channel and the tidal prism through this cross section based on estuarine data along the US coast and the Waddenzee, the Netherlands. These relationships assume morphodynamic equilibrium, but do not explain the origin of the equilibrium and its sensitivity to changing conditions. Only few researches were carried out in the laboratory investigating equilibrium conditions. Tambroni et al (2005) investigated equilibrium conditions of the width-averaged longitudinal bed profile for both a rectangular basin and a basin with an exponentially decaying width. They found good agreement with earlier results from a width-averaged 1D mathematical model by Lanzoni and Seminara (2002).

With respect to pattern formation, Schuttelaars and De Swart (1999), Seminara and Tubino (2001), Schramkowski et al (2002) and Van Leeuwen and De Swart (2004) describe initial channel/shoal formation in a highly schematized tidal environment. They found that dominant morphological wavelengths are determined by an approximate balance between the destabilizing divergence of the sediment transport field and the stabilizing divergence of sediment fluxes induced by the presence of a bed slope. Typical feature of the initially developing bars is that they can be related to the (short) basin length, the embayment width and the relative importance of diffusive and advective transports [Van Leeuwen and De Swart (2004)], or the tidal excursion length implicitly taking into account the impact of friction and depth [Schramkowski et al (2002)]. Also, the sensitivity to diffusive or advective transports was investigated by Van Leeuwen and De Swart (2004). As the bars grow in height, the effect of the bed slope on the bed load transport will be more pronounced. Additionally, there will be a more impact of non-linear interactions between the bed and the hydrodynamics. Coevelt et al. (2003) concluded, based on 3D hydrodynamic calculations, that secondary flow (so called "spiral flow") in bends is of minor importance. Schramkowski et al (2004) came to similar conclusions stating that a 2D flow description suffices for long-term bar formation so that detailed 3D model descriptions would not be necessary. The latter research pointed additionally to different types of equilibrium conditions for finite amplitude morphological waves (stable, periodically stable and unstable). Although they investigated the possible stability of specific morphological features, they did not conclude on a possible final equilibrium state of the channel/shoal system. Long-term, 2D, simulations by Hibma (2003b, 2003c) show stable patterns that have a characteristic length scale with an order of magnitude of the tidal excursion, although they are found to be 10% longer than the bars that emerge initially. This is attributed to the higher velocities in the channels between the bars.

Equilibrium of the longitudinal width-averaged bed level was investigated extensively using 1 dimensional (1D) models, which describe the propagation of a tidal wave into the estuary and relate this to morphodynamic conditions like equilibrium or ebb or flood dominance. For example, Boon and Byrne (1981) and Friedrichs and Aubrey (1988), pointed to the importance of the storage of tidal water on shoals and the interaction of the M_2 and M_4 tidal constituents (the latter also by Schuttelaars and De Swart (1996 and 2000) and by Hibma (2003a)). Friedrichs and Aubrey (1996) investigated separated effects of tides and waves. Schuttelaars and De Swart (1996 and 2000) pointed to the relative importance of advective and diffusive transports and of the basin length itself. Finally, Lanzoni and Seminara (2002) stressed the impact of the (funnel) shape of the estuary.

2.1.2 Aim of the study

Based on the foregoing, two main considerations are put forward. Firstly, pattern formation and longitudinal equilibrium profiles have been investigated separately. By definition, 1D models do not take into account a laterally non-uniform velocity distribution due to shoals and channels present in a cross-section. Considering that sediment transport depends faster than linear on the velocity, it is expected that channel/shoal patterns will at some stage impact on bed evolution and the (assumed) equilibrium conditions. Only Hibma (2003b) investigated both aspects in parallel using a 2D process-based model, although the focus of that study was on the pattern formation. Secondly, the empirical relationship derived by O'Brien (1969), Jarrett (1976) and Eysink (1990) as well as the 1D models suggest that equilibrium is present, despite the fact that especially the sediment transport is a highly non-linear processes.

The main aim of the current research is to investigate the characteristics of morphodynamic evolution in a tidal embayment, based on the application of a 2D numerical, process-based model. This allows for a prediction of the evolution based on the inclusion of detailed hydrodynamic and morphodynamic processes. For example, it combines processes like pattern formation and profile evolution including their mutual dependency and different time scales and spatial scales. Simplifications are made in the model to make comparison with earlier research possible and to be able to analyze the results in a relatively straightforward way. These simplifications are mainly related to the sediment transport formulation and the model configuration. In order to validate the model results, emphasis is put on comparison with data from the Western Scheldt estuary, see Figure 2.1, so that the model configurations are based on dimensions similar to this estuary.

The main aim of the study is systematically addressed by a number of more specific research questions. First question is related to the evolution characteristics of the longitudinal profile in a 1D schematization, so that evolution can be studied for a relatively simple configuration. Second question considers the evolution and behavior of different 2D schematizations, including an analysis of the different time scales and spatial scales in the evolution. Thirdly, it is questioned to what extent the 2D model results compare with empirical relationships and data from the Western Scheldt estuary. Final research question is in how far 1D and 2D results coincide and what the reason might be for possible differences. In the analysis special emphasis is put on a discussion of the performance of the applied morphological update scheme.

The following sections will first describe the formulation of the hydrodynamic and morphodynamic model, which is basically the same for the 1D and 2D schematizations. Secondly, the model results are presented for the 1D schematization, the 2D schematization, their comparison with each other and with empirical relationships and data from the Western Scheldt estuary. Finally, the modeling approach and results will be discussed in more detail.

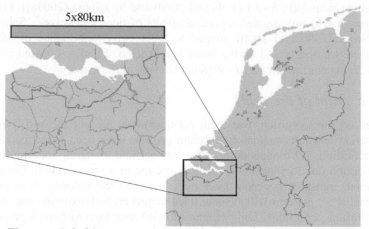

Figure 2.1 *Western Scheldt estuary, the Netherlands and Belgium (source : Rijkswaterstaat Adviesdienst Geoinformatie en ICT)*

2.2 Model geometry and configuration

The hydrodynamic and morphodynamic model applied (Delft3D) is described in detail in Section 1.4. In this section only details are given as far as these deviate from the standard model settings and parameter values.

The configuration of the model is highly schematized and consists of a rectangular box that has an open boundary at the seaward end, see Figure 2.2. Thus, no flow or transport is allowed across the sides and at the landward end. At these locations a no-slip condition applies. The reason why a rectangular configuration was chosen is twofold. Firstly, it makes comparison possible with earlier research carried out by Schuttelaars and De Swart (1996, 1999, 2000), Seminara and Tubino (2001), Schramkowski et al (2002, 2004) and Hibma (2003b) who all used rectangular basins or assumed that no significant changes of the basin width occurred over a typical length scale of the developing bars. The second reason is that a basin with an exponentially varying width or a real geometry would complicate the analysis of model results, especially concerning the hydrodynamic behavior.

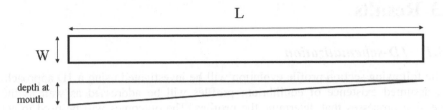

Figure 2.2. *Map (upper part) and longitudinal cross section (lower part) of standard model configuration where W = basin width and L= basin length.*

The length (L) of the embayment for the 2D schematization is 80 km and the width (W) has a value of 2.5 km. This size was chosen because it somewhat resembles the size of the Western Scheldt estuary in the Netherlands so that comparison to measured data would be possible, see Figure 2.1. The Western Scheldt is particularly suitable for comparison with this model since river discharges are very small compared with the tidal prism (<1%), so that the morphodynamics are tide dominated. The current model deviates from the model described by Hibma et al. (2003b) in the sense that erosion of dry cells is allowed, that the 'online' approach is applied with a fixed morphological factor instead of tidal residual transports with a MF depending on courant conditions, that a different formulation of the bed slope effect is applied and that longer runs are made (i.e. 800 years instead of 120 years).

In order to investigate the sensitivity of the model to different basin lengths and widths, additional 2D runs were performed for a length of 20 km and widths of 1.25 km and 5 km. It is noted that 20 km basins correspond to a short basin having a length that is small compared to the tidal wavelength. The 80 km basins approach values of ¼ of the tidal wavelength. This induces resonance resulting in observed increasing amplitude (up to 5 m.) towards the head.

The grid size of the 1D model schematization is uniform and 62.5m (lateral) x 125m (longitudinal) in size. The grid size of the 2D model is uniform and is 62.5m (lateral) x 125m (longitudinal) m thus allowing a detailed description of the growth of bars that are typically 1-2 orders of magnitude larger than the grid size. The boundary at the seaward end is defined by a varying water level with amplitude (A) of 1.75m and a period (T) of 12 hours slightly deviating by 42 minutes from the M_2 tidal period. The sediment transport at this boundary was defined by an equilibrium transport corresponding to the value of the velocities at the boundary. The boundary thus allows for net sediment fluxes over a tide. The grain size of the sediment is uniform and has a value (D) of 240 μm. Courant conditions required a hydrodynamic time step of approximately 1 minute. Standard value for bed slope factor α_{bn} was set to 5. Although Ikeda (1982) and Ikeda and Aseada (1983) suggest a value of 1.5 based on small-scale experiments, this lead to unrealistic channel profiles.

2.3 Results

2.3.1 1D-schematization

In the following section profile evolution will be investigated using a 1D approach. The assumed existence of equilibrium profiles will be addressed as well as the governing processes that determine the profiles. The outcomes will be used to act later as initial conditions for the 2D schematization.

1D model runs were carried out using the same settings as for the 2D model. Figure 2.3 (a) shows the profile evolution starting from an initially horizontal bed level at 10m-MSL (Mean Sea Level) along an 80 km long basin. Halfway the basin a front migrates towards the head, whereas at the same time the bed level more seaward deepens considerably. After 8000 years the model results show a profile sloping from about 30m-MSL at the mouth towards approximately 2m+MSL near the head. Sedimentation takes place near the head although this takes place at a decreasingly smaller rate. Figure 2.3 (b) shows similar profile evolution from a different and shallower initial profile, i.e. sloping from 15m-MSL at the mouth towards MSL at the head.

After 8000 years both profiles are approximately similar apart from the seaward shift for the shallow profile that shows more sedimentation near the head. Figure 2.3 (b) includes the profile of Figure 2.3 (a) fit by a sea ward shift. The higher sedimentation rates in case of the shallow initial profile are attributed to the shallowness of the basin. The shallower a basin becomes, the more ebb flows and ebb wave propagation are hampered so that ebb duration increases. This leads to smaller flood duration, higher flood velocities and more sediment being transported landward during flood. Figure 2.3 (c) shows that even after 8000 years no equilibrium was found in terms of a constant bed level at the mouth and halfway the basin, although evolution shows strongly asymptotic behavior.

Figure 2.3 (d) shows the results of a sensitivity analysis carried out for different values of the MF (ie.1, 10 and 400 after 20 years and 10 and 400 after 100 and 200 years) for different points in time for the basin with the horizontal bed at 10m-MSL. The analysis shows that different values for the MF result in similar behavior of the profile with only (very) small differences even on the longer term. The pronounced peaks in the Figure 2.3 (d) especially at 100 and 200 years are related to the local and temporal conditions of the steep front of the bed migrating towards the head. Bed level differences for different MF's reduce again when the steep front passes along. Sensitivity analysis was also carried out using a 2000 years bed level generated by a MF of 400 as initial bed and applying different values for the MF (i.e. 1 and 400, not shown). This showed even smaller differences than starting from a linear or horizontal bed.

Various researchers investigated 1D profile evolution in a tidal embayment using a model description similar to the current approach, i.e. water motion by the depth integrated shallow water equations in combination with a sediment transport formulation. Van Dongeren and De Vriend (1994) and Schuttelaars and De Swart (1996, 2000) found a linearly sloping bed level for a basin that is short compared to the tidal wave length (just like in the current research), despite the fact that they used slightly different formulations for the sediment transport. Both imposed a zero sediment transport at the head. Schuttelaars and De Swart (1996, 2000) based their equilibrium profiles on the criterion of vanishing tidally averaged transports, which excluded the process of drying and flooding during the tide. Van Dongeren and De Vriend (1994) argue that the evolution towards the equilibrium is not obvious, because the boundary condition of zero sediment transport at the head would lead to continuous sedimentation and the creation of intertidal area at the head, especially when conditions are considered that are not tidally averaged. Equilibrium would finally develop in a balance between the basins' infilling capacity and the exporting effect due to the creation of intertidal area. The latter effect is extensively described by Friedrichs and Aubrey (1988) and is caused by the fact that intertidal area hampers the propagation of the flood wave, so that the flood period is elongated and ebb velocities and sediment export increase.

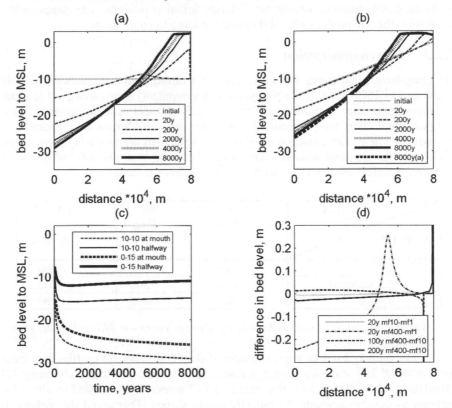

Figure 2.3 *1D model results (a) profile evolution from horizontal flat bed at 10m-MSL; (b) profile evolution from linearly sloping bed from 15-MSL to MSL; (c) bed level development over time at mouth and halfway the basin for the different initial conditions; (d) differences in bed level for different morphological factors at different points in time.*

Lanzoni and Seminara (2002) included a formulation for wetting and drying of intertidal area and solved the set of shallow water equations in a similar numerical approach as in the current research, although they assumed a different basin geometry with a landward exponentially decaying shape, included both bed load and suspended load transport, applied a different description for drying and flooding and did not extend their duration more than 300 years. Their results (for a 30 km basin) show a concave profile with asymptotically vanishing tide-residual sediment transports and the continuous formation of intertidal area near the head. It would be expected that longer model runs would reveal the formation of more intertidal area near the head, similar to the current 1D model outcomes.

The short literature overview indicates that a proper formulation of the boundary including drying and flooding of cells at the head is essential. The current model applies an approach that includes drying and flooding of the intertidal area developing near the head. The results show a continuous sedimentation at the head. Furthermore, shallower initial conditions will lead to more sedimentation. The sedimentation process only decreases because the deepening of the more seaward-located profile leads to lower transport rates towards the head. The shape of the evolving profile remains similar for different initial conditions. This supports the suggestion that the profiles after 8000 years are close to equilibrium.

2.3.2   2D-schematization

2D Long-term calculations (up to 4 years of hydrodynamic calculations with an equivalent of 1600 years of morphodynamic development) were carried out in order to investigate the evolution of the specific model configurations. Since computational effort took typically 2 weeks per run, only 8 runs are elaborated in this article. These configurations are given in Table 2. In the following sections the model results are analyzed. An animation shows an example of 800 years of morphodynamic evolution (Van der Wegen and Roelvnik, 2008).

Length (km) ⟍ Width (km)	20 short	80 long deep	80 long shallow
1.25 narrow	1600	800	800
2.5 standard	1600	800	800
5 wide	800	none	400

Table 2 *Overview of morphological duration in years for different model runs*

The authors used an approximation of the 8000 years 1D profiles as a basis for the initial bed level for further research. The reason was that preliminary 2D calculations showed that long run durations (~2 weeks) were needed to allow for patterns to develop towards dynamically stable shapes. This urged the authors to find 2D initial conditions close to eventual equilibrium profiles. The profiles were linearized starting from MSL at the head to 8m-MSL for the short basin and 34m-MSL for the long basin. However, preliminary calculations showed (see section 3) that patterns hardly developed in this long basin. Therefore, and because width-averaged bed levels in the Western Scheldt estuary are typically in the order of 10-15m-MSL, the initial 2D bed level was taken as linearly varying from 15m-MSL at the mouth to MSL at the head, which is referred to as the long, shallow basin. For all 2D configurations the initial bed was disturbed randomly with values of maximum 5% of the local water depth in order to trigger channel and shoal development.

2.3.3 Pattern formation

Figure 2.4 presents an overview of the pattern formation after 15, 100 and 800 years for a 2.5 km wide, long, shallow basin. Also, the attached * animation shows 800 years of morphodynamic evolution for the 2.5 wide and 80 km long basin. Initially the model shows major bar growth in the relatively shallow part near the head including relatively small bars. Seminara and Tubino (2001), Schuttelaars and De Swart (1999), Schramkowski et al (2002) and Dronkers (2005) extensively describe the prevailing processes in bar development for the linear domain where the bars themselves do not yet significantly influence the velocity field. Local tidal velocities and bed slope effects on the sediment transport govern the dominant length scale of these bars and their growth rates.

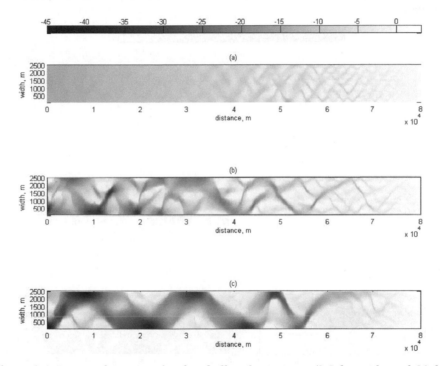

Figure 2.4 *Pattern formation for the shallow basin in m (2.5 km wide and 80 km long), (a) after 15 years, (b) after 100 years, (c) after 800 years. Not on scale.*

Later on and more seaward a front of the channel-shoal pattern develops towards the deeper parts of the basin. The front does not originate from instabilities caused by local tidal velocities like in the more shallow parts of the estuary, but it emerges from high ebb velocities elongating the channels while bars develop at the seaward end of these channels. Finally, some crests of the bars enter the intertidal domain, which makes them less subject to hydrodynamic processes. Although migration, growth and erosion are still going on, the bar position and shape become less dynamic. Seaward bar migration still takes place after 100 years, but the migration rate decreases exponentially with time. It becomes probably finally related to the (slow) adaptation rate of the longitudinal profile.

Figure 2.5 shows results after 800 years for both the short basin and the long shallow basin with varying widths. It can be clearly seen that relatively long and narrow basins generate an alternating bar pattern separated by a sine-shaped channel. Near the mouth the alternate bars are sub tidal, but more landward the bars have crests in the intertidal and supra tidal domain. Still after 800 years these bars have a tendency to migrate seaward, with a typical rate of 1 km/100 year. The meandering channel significantly deepens at the sides of the basin. Occasionally, the channel includes the characteristics of a combined ebb and flood channel separated by a small shoal in between. The wider the channel bed, the more it becomes subject to distortions, and bar development. It is believed that in case of wider basins the sine-shaped channel becomes so wide that the bars in the channel disturb the channel hydrodynamics in such a way that a dynamic multiple channel system will develop.

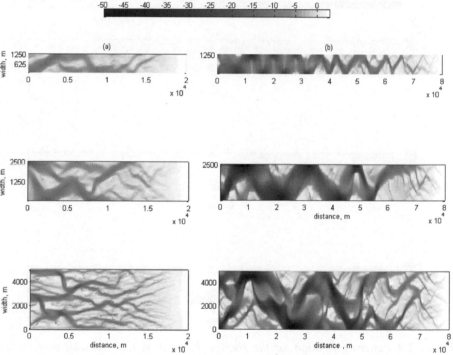

Figure 2.5 *Not on scale. Pattern formation in m for (a) a 20 km long basin with varying widths (presented results after 1600, 400 and 400 years respectively), (b) a 80 km long, shallow, basin with varying widths (presented results after 400, 1600 and 400 years respectively).*

The pattern of relatively short and wide basins is more determined by free behavior of the bars in the sense that their development seems not restricted by the banks of the embayment. After 800 years the channels keep showing a meandering behavior and no single channel dominates the pattern formation. This also holds for the shallower parts of relatively narrow basins near the head.

The long deep basin (not shown in the figure) shows similar behavior, although intertidal area only develops near the head and the front of the pattern formation does not reach the mouth, not even after 1600 years.

Based on a 2D model Schramkowski et al (2004) extend their work of 2002 to the non-linear regime and found conditions for static and dynamic (pulsing) morphodynamic equilibria of patterns. These equilibria were related to narrow (with a width small compared to the tidal excursion length) and relatively deep basins and refer to an alternating bar pattern also found in the current model. For wider and shallower basins, only time-dependent states could be found. This agrees well with the current model results.

2.3.4 Morphological wavelength

Model results show a range of characteristic bed forms. Yalin and Da Silva (1992) suggest a relationship between the spacing of the bars and the basin width in a situation of alternating bar forms based on an inventory of empirical research:

$$L_B = 6B \qquad (2.1)$$

where
L_B length of bar spacing, m
B   basin width, m

The relationship was derived and validated for relatively small values of the width to depth ratio (10< width/depth <100) and for fluvial conditions only.

Contrary to equation (2.1) the model results suggest that a characteristic morphological length scale (defined as the characteristic morphological feature that can be observed) is not solely and linearly a function of the basin width. We suggest an equation which relates the morphological wave length to the tidal excursion, the width-averaged depth and the basin width as follows:

$$L_{mw} = \delta \frac{\sqrt{\alpha_{bn}}}{c_f} \frac{\sqrt{BE\bar{h}}}{T\sqrt{g}} \qquad (2.2)$$

or, in other terms,

$$L_{mw} = \delta \frac{\sqrt{\alpha_{bn}}}{c_f} \frac{P}{T\sqrt{Bg}} \qquad (2.3)$$

where

L_{mw} morphological wavelength, m
δ coefficient, m^{-1}
B   basin width, m
E   tidal excursion defined by P/A, m
P   tidal prism (amount of water flowing through a cross-section from Low Water Slack to High Water Slack at a certain location), m^3
A   minimum cross section below MSL, m^2
$\bar{h}$ minimum width-averaged water level below MSL, defined by A/B, m
T   tidal period, s

The equation was derived by curve fitting and systematically varying the model parameters basin width (B), the tidal prism (P) and the minimum cross section below MSL (A). The tidal excursion (E) was defined by (P/A) and the mean water depth ($\overline{h}$) by (A/B), thus representing the minimum width-averaged water level below MSL. The model parameters were investigated for a root, linear and quadratic relationship (both in the nominator and the denominator) with the morphological wavelength.

Figure 2.6 (b) plots a comparison of model results observations and results from equation (2.3) for a value of 0.3 for (δ). For dominant alternate bar patterns, the morphological length scale was determined by multiplying the distance between a channel sinus 'crest' and a 'trough' by two. For dominant multi-channel beds, the length of the bars was supposed to determine the morphological length scale. Data from the Western Scheldt resemble to the model results of the long basin only to some extent. This is attributed to the meandering shape the Western Scheldt itself has, see Figure 2.1, the presence of non-erodible layers as well as model idealizations like the formulation of sediment transport and bed slope effects. The discussion of the current chapter will discuss the characteristics of equation (2.3) in more detail.

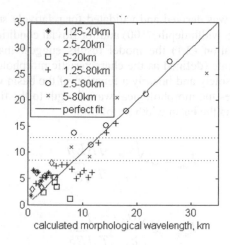

Figure 2.6 *Morphological wavelength calculated by equation (2.3) versus the morphological wavelength observed in model results; range of data from the Western Scheldt estuary are represented by the intercepted lines.*

2.3.5   Longitudinal profile

For the short, 2.5 km wide basin, Figure 2.7 (a) shows longitudinal profiles of the width-averaged bed level for different points in time. The initial, linear profile apparently does not fit the hydrodynamic conditions, since a more concave profile develops over time. At the head sediment settles slowly, although the rate seems to decrease exponentially with time. The 1D model results, starting from the same initial bed level, show similar behavior although the profile does not show disturbances from pattern formation. Figure 2.7 (b) shows comparative 2D results for different widths of the basin. Results originating from a different randomly distributed initial bed (not shown) show a range of similar deviations of the profile near the mouth. In Figure 2.7 (c), the shallow, long basin shows an increasing deepening of the concave profile. It is remarkable that the bed level at the mouth holds the value of 15m-MSL, although the bed levels in the basin itself continue to decrease. The 1D model results show considerable sedimentation at the head which is even more pronounced than the results of the 2D model (compare Figure 2.3 (b) and Figure 2.7 (c)). The 2D model transports the excess sediment mainly on intertidal shoals generated along the complete basin. This allows for less sedimentation near the head.

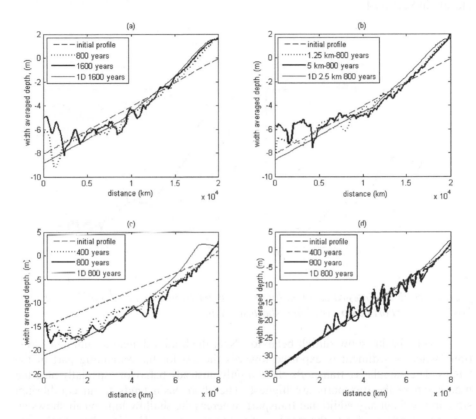

Figure 2.7 *Longitudinal profile of the width-averaged depth (a) for different points in time for a short 2.5 km wide basin; (b) for short basin width different widths after 800 years of morphodynamic calculation; (c) for different points in time for the shallow, long basin; (d) for different points in time for the deep, long basin with width of 2.5 km.*

Figure 2.7 (d) shows results for the deep, long basin with only minor sedimentation at the head, because of the deeper initial profile. The front of the pattern formation towards the sea can still be recognized, indicating that this is still an ongoing process after 800 years. Apparently, the large depth decreases tide residual transports and increases the morphodynamic time scale significantly.

2.3.6   Cumulative transports

Figure 2.8 shows width-integrated cumulative sediment transports through different cross-sections for the short and shallow, long basins, both 2.5 km wide. It is noted that the lines in the Figure 2.show long-term behavior, whereas a look on a more detailed time scale would reveal tidal fluctuations of the sediment transport. It appears that the tide residual transports have their own, relatively large time scale, although they are smaller than typical transports during the tide. This is despite the application of the high (~400) morphological factor. This will be discussed more closely in Section 4.

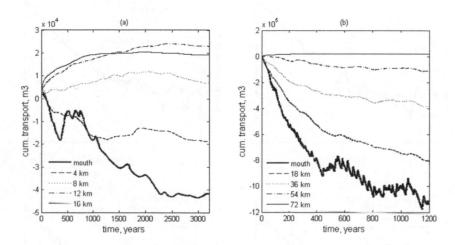

Figure 2.8 *Width-integrated cumulative sediment transport in m³ for 2.5 km wide basin for (a) short basin ; (b) long, shallow basin.*

Both basins show similar behavior. Near the head sediment accumulates over time, whereas sediment is exported towards the sea for the remaining part of the basins. The cumulative transports show highly unsteady behavior especially near the mouth where the transports are highest. The short basin tends to an equilibrium situation without any additional transport, whereas the shallow long basin shows an exporting tendency due to the surplus of sediment present. The deep long basin (not shown in the figure) does only export small quantities of residual sediment at the mouth.

2.3.7   Comparison to empirical relationships

One of the possibilities to investigate a possible equilibrium in the model is to compare the results with empirically derived relationships. O'Brien (1969) derived an empirical relationship between minimum cross-sectional area and the tidal prism through the cross-section. The tidal prism is defined as the amount of water flowing through a cross-section between low water slack and high water slack. Jarrett (1976) re-analyzed the data of O'Brien and added data published by other authors. Data were derived from 108 inlets on the US Gulf coast, the US Pacific coast and the US

Atlantic coast. The equation derived for basins that had two jetties at the mouth (for the prevention of sedimentation caused by long shore drift) reads as follows:

$$A = 7.490 * 10^{-4} * P^{0.86}$$ (2.4)

Where

A   cross sectional area below MSL, m^2
P   tidal prism, m^3

Jarrett also derived a relationship for non-jettied or single jettied basins, but these relationships match the model results less. The assumed reason for this is that the two jettied basins more closely match the rectangular model basin, allowing no influence of wave induced sediment transport at the mouth.

Figure 2.9 (a) and (b) show a comparison of the model results and equation (2.4). With time, all basins evolve towards a line similar to the Jarrett relation. However all model results over predict the cross-section with respect to equation (2.4). This is in line with observations that the model generates relative deep channels. The slope of the line (represented by the 0.86 coefficient in equation (2.4)) fits the short basins well and model results suggest a steeper slope for the long shallow basin. Both the impact of friction and the behavior of the tide could be held responsible for this. The short basin is short compared to the tidal wave length, whereas the long basin will face more impact of friction and the basin length approaches a value of a quarter of the tidal wave which induces resonant behavior.

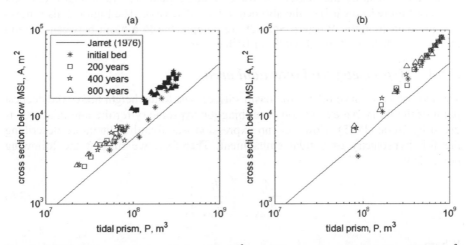

Figure 2.9 *Relation between tidal prism (m^3) and the cross-section below MSL (m^2) for different points in time. Per time point 7 to 8 locations in longitudinal direction (equidistance) are plotted on a logarithmic scale. The line represents the relation by Jarrett (1976) for jettied inlets, see also equation (2.4). (a) Short, 1.25 km wide basin (open markers) and short 5 km wide basin (filled markers); (b) long, shallow 2.5 km wide basin.*

Comparable to the investigations by O'Brien and Jarrett, based on empirical research Eysink (1990) suggests a relation between a characteristic tidal volume and the volume that the channels hold as follows:

$$V_c = c_c P^{\beta} \tag{2.5}$$

where

V_c	channel volume below MSL, m^3
c_c	empirical coefficient
P	tidal prism, m^3
β	empirical power coefficient

For different basins in the Netherlands, Eysink calibrated the coefficient c_c to value ranging from $80*10^{-6}$ (Waddenzee, large intertidal area) to $65*10^{-6}$ (Western Scheldt, small intertidal area). β was kept constant at 1.5. Figure 2.10 shows a comparison of different model results, equation (2.5) (with $c_c = 65*10^{-6}$ and $\beta = 1.5$) and measured values from Eysink (1990). Development of the model results over time is only limited so that only end model results are plotted. The Figure 2.shows that the basin width has an important impact on the model results. Relatively wide and short basins have a smaller channel volume (or larger intertidal area) for the same tidal prism. This is also apparent from Figure 2.4 . Furthermore the model results show a steeper relationship than suggested by Eysink. For example, a best fit of equation (2.5) to the model results of an 80 km long 5 km wide basin resulted in a value of 10^{-8} for coefficient c_c and a value of 1.85 for β. This can be related to the fact that the model leads to a larger increase of intertidal area more landward. An explanation could be the neglect of waves eroding tidal flats (although this effect decreases more landward) or the absence of river discharge. The Figure 2.also shows that the data from the Western Scheldt are comparable to a basin between 2.5 and 5km wide, which is a typical width for both estuaries.

2.3.8   Hypsometry and intertidal area

The hypsometric curve relates the basin surface area to the height above the deepest point of the basin. An expression describing the hypsometry for the intertidal area is given by Boon (1975). However, no expression was found in literature describing the full hypsometry of a tidal embayment. Therefore, we suggest the following power law

$$\frac{A_r}{A_{max}} = \left(\frac{h_r}{h_{max}} \right)^{\alpha} \tag{2.6}$$

Where

A_r	basin area at certain height above minimum depth, m^2
A_{max}	area at maximum depth, m^2
h_r	height above minimum depth, m
h_{max}	maximum depth, m
α	coefficient, -

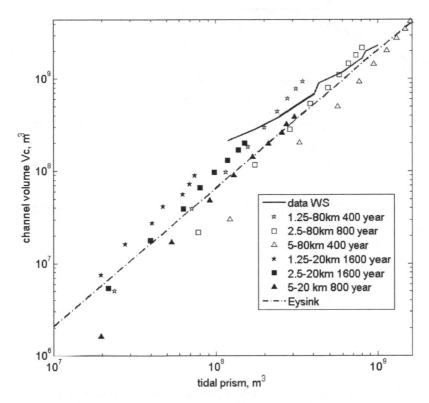

Figure 2.10 *Relation between channel volume below MSL and tidal prism. Filled markers represent the short basin for different widths; clear markers denote the long shallow basins. Comparison is made to measured values of the Western Scheldt (from Eysink 1990). Also the Eysink relation (equation (2.5)) is shown.*

Figure 2.11 (a) shows the development of the hypsometry for four points in time and an exponential fit by equation (2.6) for a short basin. No major change of the hypsometric curve takes place over time suggesting that a certain type of equilibrium is present. Figure 2.11 (c) and (d) show the hypsometry for different widths for the short (after 1600 years) and long (after 800 years) basin. The short basin shows a more convex profile than the long basin. Convex hypsometry is associated with relatively deep channels covering a relatively small area of the basin. Figure 2.11 (b) presents an overview of the values for α for different sizes of the basin. It shows that the short basins are more convex than the long basins although this difference reduces for larger widths. Although the time period for the short basin is twice the time period of the longer basin, this does not significantly change the value of 'α' considering the fact that the curves do not deviate much in time (see Figure 2.11 (a)). Comparing model results by data from the Western Scheldt in Figure 2.11 shows that the latter is more convex than the model results.

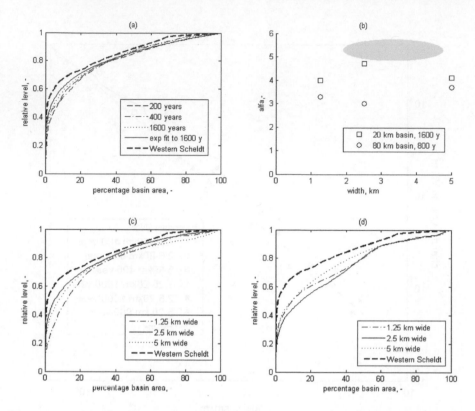

Figure 2.11 *(a) Hypsometry for different points in time for a short basin, 2.5 km wide, (b) α-values for different widths and short and long shallow basin, (c) Hypsometry of short basin after 1600 years for different widths, (d) Hypsometry of long shallow basin after 800 years for different widths. For (a), (c), (d) Western Scheldt data are bold. For (c) WS data are represented by the shaded area.*

Results of the percentage of intertidal area compared to the basin area at high water are shown in Figure 2.12 . It can be seen that the short basins show larger percentages of intertidal area. This corresponds to the more convex hypsometries that are associated with relatively deep channels covering a relatively small basin area. In addition, the Figure 2.shows that the amount of intertidal area is rising with time (except in case of a 5-80 km basin) and that there is no indication for an asymptotic behavior towards equilibrium. This can be associated with the continuing sedimentation at the head, see also Figure 2.7. Furthermore it can be concluded from the Figure 2.that the percentage of intertidal area is rising with increasing basin width, although this effect is less pronounced in case of longer calculation periods. The long deep basin has considerably less intertidal area. Data for the Western Scheldt (with a width ranging from 3-5 km in the first 50 km from the mouth) indicate values between the deep and the shallow long basins.

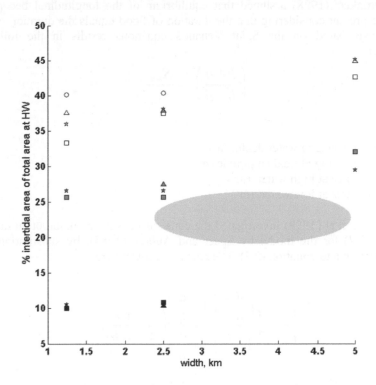

Figure 2.12 *Relation between basin width and % of intertidal area with respect to the area at high water. Clear markers represent the short basin, shaded markers the long, shallow basin and filled markers the long deep basin.* ■ *denotes 200 year;* * *denotes 400 year;* ▲ *denotes 800 year;* ● *denotes 1600 year. The shaded area shows a range of values found for the Western Scheldt over the last decade for the first 50 km. These values are not directly related to a width.*

2.3.9 Ebb and flood dominance

In the following section the effect of tidal distortion on flood or ebb dominance and the morphodynamic process will be elaborated.

Based on a 1D model, Speer and Aubrey (1985) distinguished two important parameters with respect to ebb or flood dominance, namely the ratio of the volume of intertidal storage and the channel volume at MSL (Vs/Vc) and the ratio of the tidal amplitude and the mean water depth (A/h). Increasing values of the latter will elongate ebb duration with respect to flood duration, because of friction and because of the difference in wave propagation (where high water propagates faster than low water), which will both elongate ebb duration. Increasing values of intertidal storage will decrease flood propagation and thus elongate flood duration. Friedrichs and Aubrey (1988) confirmed the model results based on measurements along the US Atlantic coast and used the model by Speer to come to predictions for ebb or flood dominance in case of adaptation of the tidal wave due to sea level rise (Friedrichs et al., 1990).

Dronkers (1998) assumed that equilibrium of the longitudinal bed profile would be present considering that the duration of flood equals the duration of ebb. An analysis based on the Saint Venant's equations results in the following expression

$$\left(\frac{h+a}{h-a}\right)^2 = \frac{S_{HW}}{S_{LW}}$$ (2.7)

with

h   width-averaged water depth, (m)
a   width-averaged tidal amplitude (m)
S_{HW} basin area at high water, (m^2)
S_{LW} basin area at low water, (m^2)

Wang et al (1999) investigated the consequences of the assumption made in equation (2.7) for the model by Speer and Aubrey (1985), by substituting their schematisation into equation (2.7). The results are given here:

$$\frac{V_s}{V_c} = \frac{8}{3}\frac{\left(\dfrac{a}{h}\right)^2}{1-\dfrac{a}{h}}\left(\frac{1+\dfrac{a}{h}}{1-\dfrac{a}{h}}\right)\left(\frac{3}{4}+\frac{1}{4}\frac{a}{h}\right)^{-1}$$ (2.8)

The equation describes a curve resembling the result of the model by Friedrichs and Aubrey (1988) and is used here mainly as an indication for flood and ebb dominance. It is stressed that Friedrichs and Aubrey used a highly schematized cross-section differing from the current model, assuming, amongst others, a fixed ratio between the channel width and the depth (value = 120), a fixed geometry in longitudinal direction, a constant offshore forcing of 0.75m and a fixed length of 7 km (which means that it is relatively short basin, contrary to the basin of the current model).

Figure 2.13 plots equation (2.8) and the current model results developing over time. The basin profiles seem to develop towards certain equilibrium asymptotically with time. Both the initial profiles show strong flood dominance, which is decreasing (at least, coming nearer towards the equilibrium line of equation (2.8)) when time proceeds. The position of the mouth is reasonably constant for t=200 years and t= 800 years and the final profiles along the basin show the same parabolic shape as equation (2.8), which is represented by the solid line. However, the position of the mouth of the short basin deviates from the rest of the basin, which is attributed to the boundary condition not allowing for overtides. The long basin does not show this deviation near the mouth. This suggests that overtides do not play a significant role probably due to the larger depths at the mouth. Further the long basin is more flood dominant near the mouth and more ebb dominant landward.

Measured values for sections of the Western Scheldt estuary derived by Wang et al (1999) correspond only roughly to the values of the long shallow basin, see the dotted area Figure 2.13 (b). Comparing the data with the model of Friedrichs and Aubrey, Wang observed that the data for the Western Scheldt did indeed correspond with the flood dominant character of certain sections, but data with relatively small values of a/h (<0.17) and Vs/Vc (<0.04) were characterized by ebb dominance in reality, which does not correspond with the flood dominant character predicted by the model.

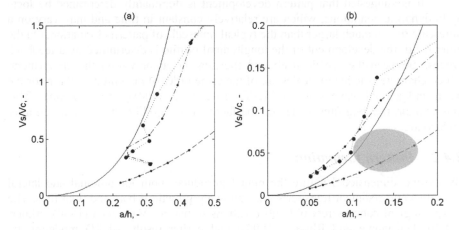

Figure 2.13 V_s/V_c versus A/h along the basin (left-down near the mouth; upper-right more landward) for different points in time.(_ _) denotes initial profile, (_._._) denotes profile after 200 years, (.......) denotes profile after 800 years; solid line represents equation (2.8) (a) 2.5 km wide short basin and (b) 2.5 km wide long, shallow basin: shaded area represents data from sections in Western Scheldt from Wang et al (1999).

2.4 Discussion

The following section addresses major issues arising from the previous sections and discusses these in more detail.

2.4.1 Two time scales

Two time scales can be distinguished in the 2D results that are related to the spatial scale of the phenomena taking place. The channel-shoal pattern with a spatial scale of O(1-10km) typically develops within the first 100 years (see Figure 2.12). After that, development still takes place, but with an increasingly slower rate related to the second time scale. Hypsometric plots that hardly develop over time, shown in Figure 2.11 , confirm this observation. The second time scale is related to the development of the width-averaged longitudinal profile with a spatial scale related to the basin width O(10-100 km). Depending on the initial bed, the longitudinal profile adaptation rate varies from centuries to millennia. For the long shallow basin, for example, the lowering of the average bed level in the basin takes place with a (slow) rate of approximately 4 meters in 1600 years.

The two distinct time scales were also reported by Tambroni et al (2005) who experimented in a laboratory flume with a long channel, closed at one end and attached to a basin with a varying water level at the other end. They investigated two model configurations, namely a rectangular basin and a basin with an exponentially decaying width. Bed patterns developed already in the first 50 tidal cycles and take the form of an alternating bar pattern, just like in the current model results. Equilibrium of the width-averaged longitudinal bed profile was reported only after 2000 tidal cycles. Just as the current model results, the resulting end profile was concave at the seaward end and slightly convex at the landward end with a shoal forming at the head.

It is suggested that pattern development is dominantly determined by local hydrodynamic conditions, which are relatively constant in time and may vary on a timescale that is much larger than the typical timescale of patterns formation. On the other hand, the development of the longitudinal profile is determined by a feedback process between the profile itself and the tidal behavior within the embayment, which characteristic length scales are of the same order of magnitude (80 km for the basin and a tidal wave length of $L=Tc=T\sqrt{(gh)}\approx12*3600\sqrt{(10*8}=400km))$. This may explain the long time scale and the fact that no equilibrium was reached after 1600 years.

2.4.2   Pattern formation

By its two dimensional nature the model introduces both longitudinal and lateral velocities and sediment transports, thus allowing growth of bars and tidal flats. The morphological patterns resemble observations in nature (Van Veen (1950), Ahnert (1960), Dalrymple and Rhodes (1995)) and earlier results of 2D modeling by Schramkowski et al (2004) and Hibma (2003c).

Equation (2.3), relates hydrodynamic and morphodynamic model parameters to a characteristic morphological wavelength. The relation of a typical morphological length to the tidal excursion is also found by Hibma et al (2003b, 2003c) and Schramkowski et al (2002). More specifically, the tidal excursion divided by a characteristic time scale (tidal period (T)) scales with a characteristic tidal velocity. The higher this velocity is, the longer distance a sand particle will travel during flood or ebb.

The linear impact of the water depth can be hydrodynamically related to the magnitude of the inertia of the water mass (see also equations (1.2) and (1.3)) just as in case of the characteristic tidal velocity described previously. Besides, there is a relatively small influence on friction via the value for c_f (see also equation (1.4)) stating that the magnitude of friction is less in deeper parts. Both effects support a larger morphological wavelength for larger ratios of inertia over friction.

A relationship between basin width and a typical morphological wavelength (the meandering wavelength in this case) was also revealed (at least for initial bar formation) both by Van Leeuwen and De Swart (2004) using a numerical model and by Yalin and Da Silva (1992) based on an inventory of empirical data of alternating bar length scales. The dependency on the basin width shows that the banks of the basin restrict the sinusoidal channel to develop longer and wider. However, the role of the basin width in the relatively short and wide basins, including multiple channels, is not clear. Although also multiple channel systems could be under influence of the basin width, it is expected that there is somehow a maximum basin width that will influence the morphological wavelength.

The dimensionless parameter (α_{bn}), see equation (1.9), influences sediment transport on a bed slope perpendicular to the flow direction. A higher value will flatten the cross-sectional profile of the basin, which allows for larger width scales of the bars and an increase of their length scales. Sensitivity analysis shows that increasing values for (α_{bn}) will increase the morphological wavelength at least by its root value.

The dimensionless parameter (c_f) has a slightly more complex relation to the morphological wavelength due to the water depth present in the denominator, see equation equation (1.4). Sensitivity analysis revealed an inversely linear relation between commonly used values of Manning's coefficient (n) ranging from 0.02 to 0.03 $\text{sm}^{-1/3}$ and the morphological wavelength.

Struiksma et al (1985) carried out a linear perturbation analysis for a similar model formulation under fluvial conditions. They derived a relation between the product of the longitudinal wave number ($k_r = 2\pi / L_{mw}$) and an adaptation length of the flow ($\lambda_w = \overline{h}/2c_f$) and the ratio of the adaptation length of the bed topography ($\lambda_s = B^2 f_s \theta /(\pi^2 \overline{h})$) and the adaptation length of the flow (λ_w), where ($f_s \theta$) represents a bed slope effect comparable to factors (α_{bn}) and (α_s) of the current model. For the range of the current model values the relationship can be represented in approximation by

$$\lambda_w \lambda_s = \frac{1}{k_r^2} \qquad (2.9)$$

This leads to the following expression for the morphological wavelength:

$$L_{mw} = \frac{B}{\pi} \frac{\sqrt{f_s \theta}}{c_f} \qquad (2.10)$$

The expression was derived for fluvial conditions with a constant mean water depth and constant velocity. Both parameters form the zero order solution, explaining their disappearance in equation (2.10). Also, a fixed transverse wave number of (π / B) was imposed across fixed banks leading to a prescribed alternating bar pattern. This alternating bar pattern excludes the possibility of multiple channels in a cross section apparent in the current model and would explain the proportionality with (B) instead of ($\sqrt{B}$) like in equation (2.2).

Further, Ikeda et al (1981) derived the following relation for river meandering length estimation based on a stability analysis of a fluvial sinuous channel with erodible banks:

$$L_{meanderlength} = \frac{\pi}{0.75} \frac{\overline{h}}{c_f} \qquad (2.11)$$

Although the relationship between alternating bar length and meandering length may not be straightforward (see also Ikeda et al (1981)), equation (2.11) shows a certain resemblance with equation (2.2). In both latter equations based on stability analysis, the velocity vanishes contrary to equation (2.2). The reasons for this might be that the velocity is not constant in longitudinal direction along the basins, due to non-linear interaction with the concave shape of the bed, friction and the tidal behavior in basins that are not short.

The current analysis shows that model parameters can be qualitatively and quantitatively related to a characteristic morphological wavelength. However, at this moment no explanation was found for the dimension of coefficient (δ). Additional research could clarify this as well as the influence of the size of the grain diameter, a grain size distribution and a different sediment transport formulation. Additionally, validation of the relationship with physical observations would need to take place.

2.4.3   Comparison of 1D and 2D longitudinal profiles

Boon and Byrne (1981), Van Dongeren and De Vriend (1994), Friedrichs and Aubrey (1996), Schuttelaars and De Swart (1996 and 2000), Lanzoni and Seminara (2002) and Hibma (2003a) showed that morphodynamic equilibrium can be reached using 1D models that include width-averaged longitudinal depth profiles.

Especially near the head, all 2D basins corresponded quite well with 1D model results. Only the shallow, long basin generated substantially more sedimentation near the head. This is attributed to the fact that (too) much sediment was available in the basin.

Just as the 1D model results, all 2D basins show a slightly concave longitudinal profile, despite the disturbance by the pattern formation (see Figure 2.7). This is remarkable, since research by Friedrichs and Aubrey (1988) suggests that the intertidal area will significantly influence the behavior of the tidal wave and the related sediment transports within the basin. However, the criterion for morphodynamic equilibrium by equal ebb and flood duration (Dronkers, 1998) may be finally fulfilled for the 1D (excluding intertidal area) and 2D (including intertidal area) configurations, despite their different dominant mechanisms. As an illustration Figure 2.13 (a) indicates that the short basin has reached certain equilibrium in terms of the relatively constant value for the (a/h) versus (Vs/Vc) relationship after 200 years at the mouth.

However, the 2D bed levels at the mouth tend to be higher for all basins except for the long deep basin. This effect does not seem to disappear over time (see Figure 2.7). Furthermore, it is observed that one relatively deep channel is persistently present at one side of the basin at the mouth and that the rest of the mouths' cross section is generally covered by a shoal. Since these phenomena seem to be an effect only taking place near the mouth in both the short and long basins, it may be possibly attributed to the numerical formulation of the boundary condition, especially where it concerns the presence of intertidal area. Apparently, for the highly schematized boundary condition (both in terms of hydrodynamics and morphodynamics), the patterns cause flow conditions that favor relatively shallow width averaged bed levels, compared to 1D situations without pattern development.

2.4.4   Tide residual sediment transport

A possible indication for reaching morphodynamic equilibrium is a disappearing tide residual sediment transport over time. Figure 2.14 shows the magnitude of the tide residual transport for the long, shallow basin for different points in time. Main transports take place in the channels. The residual transports decrease along the basin over time although the major transports at the mouth still differ at least an order of magnitude with the rest of the basin. The direction of the residual transports becomes clear in Figure 2.15 showing the bathymetry and the residual transport made dimensionless per cell by the RMS value of the sediment transport over one tide per cell for the long shallow basin after 800 years. The ebb dominant transport is maintained on the shoals and at the landward side of the shoals and that flood dominance occurs at the seaward side of the shoals and in the channels.

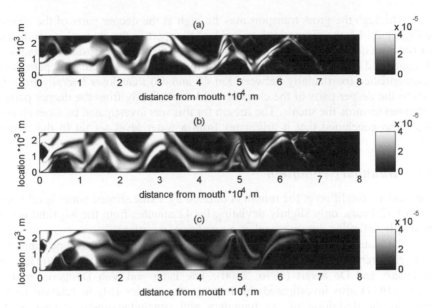

Figure 2.14 *Magnitude (in m³/sm) of tide-residual transport for long shallow basin after (a) 200, (b) 400 and (c) 800 years.*

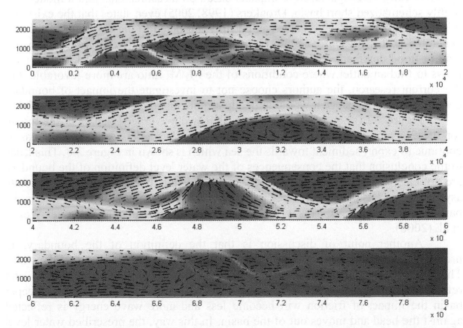

Figure 2.15 *Map of (in surface) bed levels and (in vectors) the ratio of the tide residual sediment transport over the local rms value of the sediment transport for the long shallow basin with a bed level after 800 years. The latter indicates the dominant direction of the local sediment transport.*

Although the gross transport may be high in the deeper parts of the channel (see Figure 2.14), the dimensionless residual transport in the channel is small apart from the area near the mouth. On the shoals the dimensionless residual transport is high compared to that in the deeper parts of the channels. A circulating pattern can be distinguished (particularly between km 42 and 49) that flows laterally from the shoals to the deeper parts of the channel and longitudinally from the deeper parts of the channel towards the shoals. The reason for this was investigated by Coevelt et al. (2003) who concluded that it originates from a water level set up in the channel bends.

2.4.5 *Boundary condition*

The boundary condition at the mouth is defined by a sine shaped water level with a period of 12 hours, only slightly deviating by 42 minutes from the M_2 tidal period. This simple definition, however, raises a number of questions.

The water level boundary was used since it makes comparison possible with earlier research by and Lanzoni and Seminara (2002), Hibma (2003a) and Schuttelaars and De Swart (1996, 2000). The latter and Van Dongeren and De Vriend (1994) also investigated the impact of overtides (M_4 in relation to M_2) present in the definition of the boundary and concluded major impact on the sediment export/import of the basin depending on the value of the overtides amplitude and its phase difference with M_2. Friedrichs and Aubrey (1988) systematically investigated these impacts based on measurements and a model of a highly schematized short basin. Dronkers (1998, 2005) even states that the existence of tidal basins in the Netherlands can be fully attributed to the character of the tide generated at the foreshore; certain ratios of the M_2/M_4 water level tide will always lead to flood dominant, estuaries, filling the basins with sediment and forcing river flows to find an outlet where conditions of the M_2/M_4 ratio are more favorable. For the current research, the authors choose not to investigate the impact of boundary condition overtides, since it would logically require numerous and lengthy 2D runs.

Figure 2.7 (a) and (b) show decreasing width-averaged depths at the mouth for the 2D model and increasing depths for the 1D model. Although, all basins continue to export sediment towards the sea which is shown in Figure 2.8. This leads to the conclusion that the consequences of the water level definition of the boundary are not clear. Although the water level boundary is purely sine-shaped, the velocity asymmetry and its phase difference with the water level plays a dominant role in the basins' import or export. This will be investigated in more detail in Van der Wegen et al. (2008).

Another point of discussion is that the definition of the boundary, in principle, consists of a signal of an incoming tidal wave and an outgoing tidal wave. The incoming part of the wave enters the basin, where it is subject to friction, and reflects against the head before going out of the basin again. In case of a deeper basin the impact of friction will become less and more wave energy is reflected against the head and moves out of the basin. In this way, the prescribed water level boundary implicitly means that, by a deepening and deforming basin during the model run, the incoming wave energy decreases. It is assumed in the current research that this effect is negligible, although further research should reveal the impact of this effect.

2.4.6 *Morphological update scheme*

The validity of the results obtained in this study depends on a proper performance of the morphological update scheme applied. Figure 2.3 (c) showed already that, for the 1D schematization, values of the morphological factor up to 400 did not significantly lead to different results compared to a morphological factor of 1. At this point it is worthwhile to take a closer look at the 2D model results and to evaluate the impact of the approach. Figure 2.16 and Figure 2.17 show a comparison of 2D model results after 20 years and 100 years for different values of the morphological factor (MF). After 20 years the difference between MF 1 and MF 10 are negligible (Figure 2.16 (b)). After 100 years a MF of 1000 (Figure 2.17 (c)) even shows similar results to a MF of 10 (Figure 2.17 (a)). The figures show that higher MF's result in larger deviations, although this is mainly due to a small phase shift of the pattern characteristics. Figure 2.18 (a) shows additionally that a high MF results in somewhat shallower channels. Figure 2.18 (b) shows that local bed levels may differ on the longer time scale for different MF's although this is mainly attributed to the pattern phase shift.

The principle of the online MF approach is clearly illustrated by Figure 2.18 (c) as a detail of Figure 2.18 (b). Apart from the initial values the bed levels show similar evolution over time, despite the fact that the higher MF results in larger deviations (proportional to the values of the MF) during the tidal cycle. As long as the bed level differences during a tidal cycle and after one time step remain low compared to the water depth, the bed level update does not significantly influence the hydrodynamics and the online approach will lead to acceptable results.

A final analysis was made on the effect of different values for the MF based on a bed level including pattern development (i.e. a bed level after 1600 years). Figure 2.19 (b) presents the difference between MF 1 and MF 400 after 10 years, which shows the relatively small differences and the fact that the major differences are located at relatively high bed slopes.

The question arises what value of the MF is acceptable. It was found that only occasionally and for local points the differences in bed level during a tidal cycle amounted up to 25 % of the water depth in case of a MF of 400. Larger values (i.e. 1000) were considered as unacceptable, because of the larger phase shift (see Figure 2.17 (d)) and the larger amount of bed level points exceeding unacceptable bed level changes.

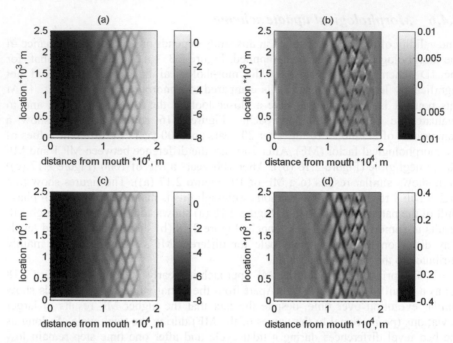

Figure 2.16 *Bed levels of short 2.5 km wide basin after 20 years; (a) results with MF of 1; (b) difference between MF 1 and MF 10; (c) results with MF 400; (d) difference between MF 1 and MF 400;*

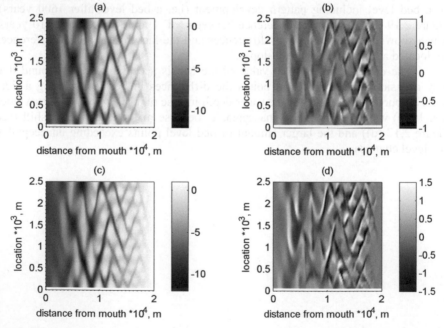

Figure 2.17 *Bed levels of short 2.5 km wide basin after 100 years; (a) results with MF of 10; (b) difference between mf10 and mf 400; (c) results with mf 1000; (d) difference between MF 10 and MF 1000.*

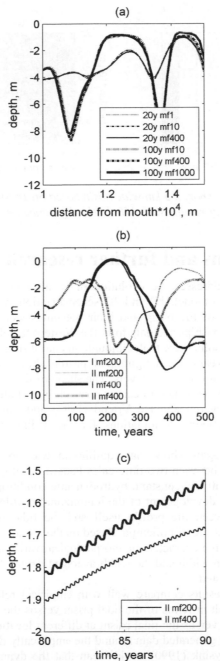

Figure 2.18 *Bed level development over time for different morphological factors for the short 2.5 km wide basin; (a) spatial plot of cross section in landward direction (b) Comparison of results of MF 200 and MF 400. Points were taken at 6 km (point I) and 11 km (point II) from the mouth and 500 m from the upper bank; (c) detail of (b) at point II;*

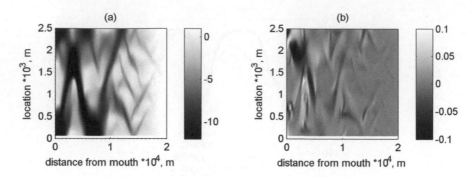

Figure 2.19 *Bed levels of short 2.5 km wide basin based on 1600 year initial bed and results after 10 years; (a) results with MF 1; (b) difference between MF 1 and MF 400;*

2.5 Conclusions and further research

The current research addresses the evolution of an alluvial embayment using a numerical, process-based model. 1D and 2D schematizations were investigated for different configurations of the basin and their outcomes were compared to each other, to empirical relationships and data from the Western Scheldt estuary.

The 1D model results show that, although no equilibrium could be reached in a timeframe of 8000 years, the longitudinal profile shows strongly asymptotic behavior for two different initial bed levels. The shape of the profiles is similar and slightly concave and can be related to earlier research by Schuttelaars and de Swart (1996, 2000), Van Dongeren and De Vriend (1994) and Lanzoni and Seminara (2002) despite their different formulations for the sediment transport and the geometry of the basin.

The 2D results again show that equilibrium was not reached. However, distinction can be made between two timescales Pattern development is dominantly determined by local, relatively constant, hydrodynamic conditions with a time scale of decades, whereas the development of the longitudinal profile is determined by a feed back process between the profile itself and the tidal behavior within the embayment with a time scale of millennia. Based on the model results a relationship could be derived between a characteristic morphological wave length and local tidal excursion, the local basin width and the width-averaged depth. Also a description of the hypsometry is suggested.

The 2D model results compare well with empirical relationships and data from the Western Scheldt in terms of the tidal prism versus the cross sectional area and the channel volume versus the tidal prism at different locations along the basin. The scatter of the model generated data around the empirically derived relationships of Jarrett (1976) and Eysink (1990) could mean that the dynamics of the system (both in the model and in reality) may fluctuate within an order of magnitude of the relationships, although additional research would be required to test this hypothesis.

Also comparison of the model results to data from the Western Scheldt estuary leads to acceptable results in terms of morphological wavelength, the description of the hypsometry and the percentage of intertidal area. This is a remarkable result especially considering the fact that, amongst others, waves, grain size distribution, presence of mud and biomass, topographic features (such as dikes and peat layers present in the Western Scheldt) and the distinction between bed load and suspended load transport mechanisms were not taken into account.

Comparing 1D results with 2D results shows similar longitudinal profiles. This is despite the fact that 2D profiles are determined by different factors determining the tidal wave propagation characteristics, which include amongst others the shoal volume and channel depth.

Major drawback of the current model is its definition of the boundary at the mouth, describing a sine shaped tide for the water level. Additionally, the rectangular shape of the basin does not correspond with exponentially decreasing estuary widths found in nature. Systematic observation of estuary geometry (for example by Savenije (2005)) shows generally an exponentially decaying width landward.

Further research (Van der Wegen et al. (2008)) will focus on a proper definition of the model boundary, the impact of erodible banks and decaying widths on long-term morphodynamic evolution as well as a more thorough analysis of the characteristic and changing hydrodynamic conditions within the basin.

3 Bank erosion and energydissipation in an idealized tidal embayment²

Abstract

The morphodynamic system in alluvial, coastal plain estuaries is complex and characterized by various time scales and spatial scales. The current chapter aims to investigate the interaction between these different scales as well as the estuarine morphodynamic evolution. Use is made of a process-based, numerical model describing 2D shallow water equations and a straightforward formulation of the sediment transport and the bed level update. This was done for an embayment with a length of 80 km on a time scale of 3200 years, with and without bank erosion effects. Special emphasis is put on analyzing the results in terms of energy dissipation.

Model results show that the basins under consideration evolve towards a state of less morphodynamic activity, which is reflected by (amongst others) relatively stable morphologic patterns and decreasing deepening and widening of the basins. Closer analysis of the tidal wave shows standing wave behavior with resonant characteristics. Under these conditions, results suggest that the basins aim for a balance between the effect of storage and the effect of fluctuating water level on wave celerity with a negligible effect of friction.

Evaluating the model results in terms of energy dissipation reflects the major processes and their timescales (pattern formation, widening and deepening). On the longer term the basin wide energy dissipation decreases at decreasingly lower rate and becomes more uniformly distributed along the basin. Analysis by an entropy based approach suggests that the forced geometry of the configurations prevents the basins from evolving towards a most probable state.

---

² An edited and slightly adapted version of this chapter was published by AGU. Copyright (2008) American Geophysical Union. To view the published open abstract, go to http://dx.doi.org and enter the DOI.

Van der Wegen, M., Z. B. Wang, H. H. G. Savenije, and J. A. Roelvink (2008), Long-term morphodynamic evolution and energy dissipation in a coastal plain, tidal embayment, J. Geophys. Res., 113, F03001, doi:10.1029/2007JF000898

3.1 Introduction

Estuaries can be defined as the area where fresh river water meets salt seawater. From an environmental point of view these are highly valuable areas. Fresh and salt-water ecosystems co-exist and interact with each other. The channel-shoal system of the estuarine bathymetry forms an environment that is tidally renewed with water and nutrients.

Because of the rich ecosystem and strategic location, estuaries have been subject to human settlement. As a result in many estuaries there is a strong impact of human interference. Examples are the construction of groins and dredging of access channels to ports. The morphodynamic system and the related ecosystems are thus artificially disturbed. Further, and on the longer term, sea level rise influences long-term behavior of the morphodynamic system and its impact on the related ecology.

3.1.1 Morphodynamic equilibrium

The concept of "morphodynamic equilibrium" refers to a certain natural and steady state of the estuarine bathymetry. This state may even be dynamic in the sense of regular or cyclic migration of shoals and channels and is subject to processes of different spatial scales and time scales interacting with each other in a process of continuous and dynamic feedback [De Vriend (1996)]. This implies that equilibrium characteristics are difficult to assess, especially concerning more geological timescales of centuries or millennia. The current research addresses the existence of assumed morphodynamic equilibrium and its related characteristics for a tidal embayment.

Generation and development of channel-shoal patterns have been studied extensively. Reynolds (1887) already investigated in laboratory flumes "the action of water to arrange loose granular material over which it may be flowing". More recent examples of numerical approaches to the phenomenon are the investigations by Schuttelaars and De Swart (1999), Seminara and Tubino (2001), Schramkowski et al. (2002) and Van Leeuwen and De Swart (2004) describing initial channel-shoal formation in highly schematized tidal environments. Schramkowski et al. (2004), Hibma et al. (2003b, 2003c), Van der Wegen et al. (2006, 2007) and Van der Wegen and Roelvink (2008) extended the research to the domain where the bed forms significantly start to influence the hydrodynamic behavior of the system, including the feedback to the morphodynamics.

On an embayment length scale, 1D model research [Van Dongeren and De Vriend (1994), Schuttelaars and De Swart (1996 and 2000) and Lanzoni and Seminara (2002)] led to equilibrium conditions of the longitudinal profile. Van der Wegen and Roelvink (2008) investigated pattern formation, the development of the longitudinal profile and their interaction in a 2D tidal domain. On the timescales considered (~1000 years), they found a relatively stable channel-shoal pattern and a continuously developing longitudinal profile. The existence of two distinct timescales was also found by Tambroni et al. (2005) based on laboratory tests. In his mobile bed physical scale models Reynolds (1887) also observed a certain steadiness in the development of the channel-shoal system, although he did not report on ongoing development of the longitudinal profile.

3.1.2   Criteria for equilibrium

This raises the question in what kind of terms an eventual equilibrium could be defined. Marciano et al. (2005) showed that channel structures generated by a 2D process-based model in a highly schematized tidal environment (similar to the model described in this article) obeyed both Horton's law for drainage network composition (Horton, 1945) and observations in the tidal environment of the Waddenzee, the Netherlands. Nevertheless, they recognized that Hortons' law cannot discriminate underlying processes and distinct network structures (Kirchner, 1993). Rinaldo et al. (1999a and 1999b) and Feola et al (1999) tested different relationships between hydrodynamic properties and morphological conditions for different tidal basin networks and conclude that many geomorphic relationships, that are valid within fluvial conditions, are site specific and do not hold across ranges of scales found in natural tidal networks.

Although the assumed presence of equilibrium is not reflected in several geomorphic relationships stemming from fluvial networks, different other criteria and dominant relationships for estuarine morphodynamic equilibrium are suggested in literature, i.e. and amongst others, between tidal prism and cross-section (Jarrett, 1967), an equal duration of ebb and flood [Dronkers (1998)], the relation between the M2 and M4 tidal constituents [Friedrichs and Aubrey (1988)], the interactions of M2, M4 and M6 tidal constituents [Van de Kreeke and Robaczewska (1993)] and the hypsometric characteristics of the estuary [Boon and Byrne (1981)]. Further, the impact of the time lag between maximum velocities and maximum water levels has been reported to be a factor of significance [Ahnert (1960), Friedrichs and Aubrey (1988), Jay (1991), Savenije (2001) and Savenije and Veling (2005), Prandle (2003)]. Lanzoni and Seminara (2002), Prandle (2003), Savenije (2001) and Savenije and Veling (2005), Toffolon et al. (2006) stress the importance of a landward exponentially decaying cross-section for the existence of equilibrium conditions. The importance of the basin length is stressed by Schuttelaars and De Swart (2000), Toffolon et al. (2006) and Savenije (2005).

The above mentioned research on equilibrium conditions in estuaries has been primarily based on an analysis of results derived from the mass and momentum equations. Singh et al. (2003) present an extensive historical overview of theories that derive relationships between hydraulic parameters and the geometry of river flow taking energy considerations as a starting point.

Rodriguez-Iturbe et al. (1992) combine the criteria of shear stress and sediment transport in their description of energy dissipation in drainage networks. They show that drainage networks in equilibrium aim at a configuration with a minimum of energy dissipation leading to fractal, tree like networks. They base their findings on three principles, i.e. 1) the principle of minimum energy dissipation in any link of the network (a local optimal condition), 2) the principle of equal energy dissipation per unit area of channel anywhere in the river network (a local optimal condition normalized by the topographical characteristics of the local area) and 3) the principle of minimum energy dissipation in the network as a whole (an optimal arrangement of the elements in the network). They suggest that when the first two principles are obeyed the third principle is met as well.

One of the main issues where river regime theory differs from theory on estuarine flow is that, in case of a river, the discharge is often given by constant upstream catchment characteristics, whereas in estuaries there is a mutual interdependency between discharge and geometry, especially considering larger timescales, see Savenije (2005; pp.10). In order to address this issue Townend (1999) and Townend and Dun (2000) focused on long-term morphodynamic behavior of an estuarine environment by considering the energy flux instead of the energy head usually applied in river regime theory. They suggest that the system leads to a condition in which the energy distribution along an estuary can be described by an exponential function. Deviations between the energy distribution function found in real estuaries and the model outcomes are attributed to impact of river discharge, sediment availability, geological characteristics and the impact of man made works (i.e. training works).

3.1.3   Aim of the study

The aim of the current research is to investigate the characteristics of 2D morphodynamic evolution of a coastal plain type of estuary with special emphasis on analyzing the model results in terms of energy dissipation.

Use will be made of a process-based, 2D model describing detailed hydrodynamic and sediment transport processes on both the scale of pattern formation and the total embayment. For an evaluation of the evolution toward equilibrium the method suggested by Rodriguez-Iturbe et al. (1992) will be applied. Amongst other methods that describe evolution and equilibrium, this description was chosen because it includes both the shear stress and the sediment transport criteria for morphodynamic equilibrium.

Two different schematized model configurations will be investigated, i.e. a configuration allowing only deepening of the profile in a rectangular embayment with fixed banks similar to that of Van der Wegen and Roelvink (2008) (referred to in the following as FBC) and a configuration allowing both deepening and widening of a rectangular embayment by allowing bank erosion (EBC). The EBC was chosen to investigate the behavior of the system allowing adaptation of geometry.

First, we address the governing equations that describe the tidal hydrodynamic and morphodynamic processes. Further, an analysis of the model results will be made in terms of energy dissipation. Finally, validation of the results is made against the Western Scheldt estuary in the Netherlands and comparison is made to equilibrium conditions suggested by an entropy based approach.

3.2   Model description

The hydrodynamic and morphodynamic model applied (Delft3D) is described in detail in section 1.4. In this section only details are given as far as these deviate from the standard model settings and parameter values.

3.2.1   Morphological factor

The previous chapter applied a value for the morphological factor of 400. Sensitivity analysis indicated that a factor of 200 is more appropriate for the current model settings, because of the large impact of the bank erosion effect. This causes relatively high sediment transports and made the basin relatively shallow at the start of the model runs. Both effects resulted in bed level updates that frequently exceeded 10% of the water depth when a morphological factor of 400 was applied.

3.2.2   Dry cell erosion and bank erosion

The reason for applying a bank erosion algorithm is twofold. Firstly in previous runs, excluding bank erosion, shoals emerged that remained dry during the rest of the run and acted as, unphysical, fixed points of the morphodynamic system. This probably happened because of water levels, dropping from high water at a particular location, were not able to reach the top of the shoal again due to small hydrodynamic adaptations during the tide. A second reason is that the geometry of a tidal basin highly affects the hydrodynamic and morphodynamic behavior. Including bank erosion in the model runs would allow for investigating its impact on emerging basin geometry and hydrodynamic conditions.

Bank erosion is a function of (amongst others) local shear stresses and sediment properties such as the angle of internal repose and cohesion. Further, just as the deepening process of the bathymetry, the widening process due to bank erosion has its own typical time scale. A particular problem in relation to numerical models is that bank erosion is a process that requires gradual erosion of banks, which often cannot be resolved by the grid of the hydrodynamic model. Research is in progress to describe bank erosion on different time scales and spatial scales. ASCE (1998a, b) and references therein provide an overview of relevant processes and modeling techniques. However, an accurate and validated approach that could be applied in the process-based numerical approach of the current research is not known to the authors.

In our model a relatively simple formulation is applied (see also Appendix I). Bank erosion is allowed by assigning the erosion that is taking place in a wet cell to the adjacent dry cell. Implicitly this means that sediment is transported from a dry cell to a wet cell so that no bed level change of the wet cell takes place. The procedure continues until the dry cell becomes wet again. The same procedure is applied for cells that become dry during the run. This can be the result of the tidal movement, or the morphological factor which locally may cause high deposition so that wet cells to become dry.

The bank erosion algorithm applied implies that the bank erosion is a direct function of morphodynamic developments taking place under water next to the dry area. The relatively large grid resolution of 100 by 200 m does not cover gradual lateral bank erosion processes. Applying smaller grid cells would describe the process more adequately, in the sense it allows for a more gradual development. Further, it implicitly also assumes that the banks consist of loosely packed sediments, so that soil properties (like, for example, compaction and moisture) or ecological impacts (Murray et al, 2007) are not considered. It is also important that the banks are not defined too high so that the gradient between the bed level in the wet cell and the bed level at the bank does not become higher than the internal angle of repose. Larger gradients introduce steep slopes and would imply that banks are cohesive, which is not the case for the current research. Roelvink et al. (2003) validate the method for a short timescale, extreme condition, namely a barrier island breach. More research would be desirable on validation of the bank erosion algorithm behavior on different grid sizes and longer time scales, but this is considered outside the scope of this study.

3.2.3   Geometry

The current research is based on an 80 km long rectangular embayment. The reason why this size was chosen is that it resembles the shape of the Western Scheldt estuary in the Netherlands. It can be classified as a coastal plain estuary following the definition of Pritchard (1952), with a meso tidal to macro tidal range following the classification by Hayes (1975).

Van der Wegen and Roelvink (2008) validated a similar model configuration based on data from the Western Scheldt estuary. One of the drawbacks of their approach was that the seaward boundary was located at the mouth of the basin and was prescribed solely by a semi-diurnal tide water level fluctuation. Although the model allowed for the generation of overtides within the model domain, describing only a semi-diurnal water level fluctuation at the estuary inlet seems inappropriate to model inlet morphodynamics. Therefore, in the current research, the model domain was extended seaward which is shown in Figure 3.1. In this way, developing overtides within the model domain account for more realistic conditions at the inlet. As a result and due to the sediment exporting character of the basin, this also allowed for the formation of an ebb tidal delta. A closer analysis of the delta characteristics will be the subject of future research.

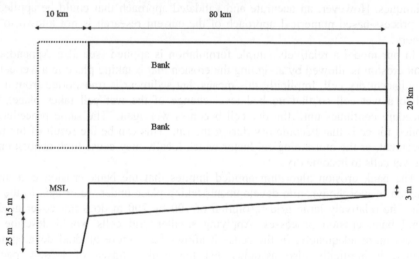

Figure 3.1 *Model configuration. The dotted line represents the seaward boundary*

The boundary condition at the seaward end (i.e. the northern, southern and western parts of the sea boundary) is described by a varying water level with an amplitude of 1.75 m and a period of 12 hours. .The boundary at the head is closed allowing no transport of water or sediment across the head. This agrees with discharges from the Scheldt River, which are negligible with respect to prevailing tidal discharges in the Western Scheldt. Furthermore, no wind or wave action, density currents or sea level rise were considered.

The width of the FBC is 2.5 km and the initial width of the EBC is 500 m. No transport of water or sediment across the banks is allowed in case of the FBC. The banks at EBC are specified at a level of MSL+3.2m, slightly larger than the maximum observed tidal water levels. This causes only gentle slopes between wet and dry cell bed levels and supports the assumed condition of loosely packed sediment. Preliminary calculations showed that, after some time, lower bank levels lead to major flooding of banks near the head, which resulted in major infilling of the embayment. This extra sediment supply acted as an extra disturbance of the system and increased the adaptation time considerably. Therefore the bank levels where set at a value that would not allow for flooding during the model runs

By running the model in a 1D mode, Van der Wegen and Roelvink (2008) showed that a bed level varying from MSL–34m at the mouth to MSL at the head resembles a morphodynamically stable profile for an embayment length of 80km. However, such a deep bed level as a starting point for a 2D configuration would mean that the 2D pattern needs a long time (~ millennia) to develop from scratch. A shallower profile is preferred, since an important aspect of the current research is to investigate the impact of pattern formation on the estuarine evolution. Additionally, comparison to the bed level of the Western Scheldt shows a much shallower basin. Therefore, the initial bed level is taken linearly sloping from MSL–15m at the mouth towards MSL at the head, so that the channel shoal system can develop within decades. Further, the initial, linear bed level was randomly disturbed by perturbations of maximum 5% of the water depth in order to provoke pattern formation.

3.3 Results

3.3.1 Fixed banks

FBC Pattern formation and longitudinal profile

Figure 3.2 shows some illustrative output from the model. Initially, the perturbations of the initial bed level grow to larger scale bed forms especially in the area near the head of the basin. The reason why the pattern development takes in that particular area is attributed to the relatively high impact of friction in the shallower part of the basin. There the flow is attracted by the deeper cells that have less friction and which interconnect and evolve into small channels. The shoals in between may even develop into intertidal area. Further in time, bed forms develop slowly towards the mouth, because the flow is disturbed by the pattern developing at the shallower part. This continues until the complete basin is covered by a channel-shoal system after approximately 100 years.

At the landward end multiple channels are distributed over the width, whereas in the area from the mouth towards about 60 km landward an alternating bar pattern is apparent. After the first decades, the main characteristics of the channel-shoal system are roughly maintained, such as bar and meander channel length scales and the alternating bar pattern. However, seaward migration of the shoals and small changes in characteristic length scales are observed. These are attributed to the deepening longitudinal profile and its impact on the tidal wave characteristics.

Due to the exporting character of the embayment an ebb tidal delta forms outside the mouth. It is characterized by a shallow wedge towards the mouth which is separated from the banks by two main channels. The channels are alternatively dominant in conveying the main tidal flow and sometimes migrate across the wedge.

Detailed analysis of the ebb delta lies not within the scope of this research and will be a subject of future research.

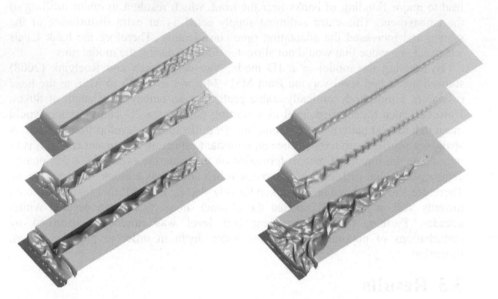

Figure 3.2 *Impression of bathymetry after 25, 400 and 3200 years for FBC (left) and for EBC (right) on a distorted scale.*

Van der Wegen and Roelvink (2008) distinguish two typical scales, which also hold for the current research. One is for the pattern formation with a timescale of decades and a spatial scale of the basin width or the water depth for the shallower parts. The other scale is related to the width averaged longitudinal profile with a timescale in the order of millennia and a spatial scale related to the basin length.

Figure 3.3 (a) shows that the width averaged longitudinal profile increases to MSL+2m at the head within the first decades. The profile at the head remains further relatively constant in time. Near the mouth the longitudinal profile continuously deepens and the profile along the basin becomes concave. This process continues but becomes gradually less pronounced. Comparison of the profiles after 1600 years and 3200 shows only little deepening with respect to the deepening compared to the profiles after 100 and 400 years. This is confirmed considering the accumulated sediment transport over different cross sections along the basin, Figure 3.4 (a). Halfway along the channel and at the mouth the embayment continuously directs sediments seaward which results in a pronounced ebb tidal delta.

FBC Tidal amplitudes

Figure 3.5 (a) and (b) show, respectively around low water and high water, the water levels at the seaward boundary compared to water levels 2 km landward from the mouth for the FBC. Small delays of low water and high water are present due to the 20 km distance the tidal wave needs to travel towards the mouth. Initially, the high water amplitude remains constant and that the low water amplitude increases.

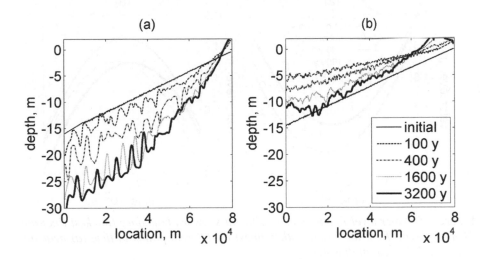

Figure 3.3 *Development of width averaged longitudinal profile over time (a) for FBC (b) for EBC*

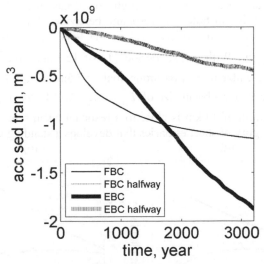

Figure 3.4 *Accumulated sediment transport through basin wide cross section (m³) over time for FBC (thin lines) and for EBC (thick lines). Solid lines show transport at the mouth and dashed lines show transport halfway (40 km landward from the mouth).*

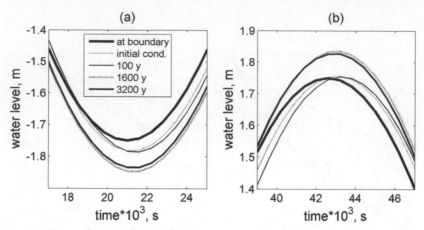

Figure 3.5 *Water level fluctuations in FBC at seaward boundary (thickest line) and 2km landward from the mouth (other lines) at different points in time (a) near low water and (b) near high water*

Figure 3.6 (a) shows that the water level amplitude decays landward in first instance, because of the high friction caused by the relatively shallow basin. However, the amplitude profile slowly evolves towards an amplitude, which linearly increases from 1.8 m at the mouth to 2.8 m at the basin head after 400 years. The linear amplification and the damping agree with Savenije (2001). Over the longer term the amplitude decays again although the decay itself becomes less. The amplitude increase is attributed to resonant behavior of the tidal wave. Under shallow water conditions the celerity (c) of the tidal wave can be approximated by $c = L_{tide} / T = \sqrt{gh}$. With a tidal period (T) of 12 hours, the tidal wave length (L_{tide}) halfway the embayment (assuming that h=7.5m) is approximately 370 km. Resonance will occur for a basin size of $L_{basin} = L_{tide} * n / 4$, *for* $n = 1, 2, 3...$ For n=1 the embayment length of 80 km is close to a resonant basin size. Hence the estuary belongs to the category of short estuaries that develops a standing wave.

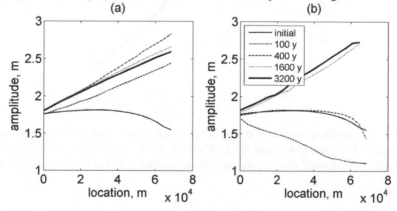

Figure 3.6 *Tidal amplitude (defined by (highest water level-lowest water level)/2) along the basin for different points in time. Data near the head are not shown since this area falls completely dry at relatively low water. (a) For FBC (b) for EBC.*

FBC Velocities, time lags and ebb/flood duration

Figure 3.7 (a) shows the development of maximum flow-width averaged ebb and flood velocities over time. Initially, ebb flows are larger than flood flows apart from the area near the head. Pattern formation within the first 100 years increases both ebb and flood flows, because the relatively high friction on the shoals stimulates water to flow through the relatively small cross sections of the channels. Over the longer term the velocities decrease again below the initial values due to the deepening of the profile and the difference between ebb and flood velocities disappears.

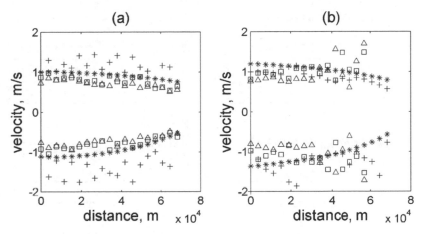

Figure 3.7 *Maximum width averaged ebb (negative values) and flood (positive values) velocities along the basin for different points in time ('*'=initial condition, '+' = 100 years, '□' = 1600 years, '△' = 3200 years). Data near the head are not shown since this area falls completely dry at relatively low water. (a) For FBC and (b) for EBC.*

Figure 3.8 (a) shows a time lag of about $1/3\,\pi$ (or about 120 minutes based on a tidal period (T) of 720 minutes) between slack water and MSL initially present at the mouth. This time lag increases to $\frac{1}{2}\,\pi$ over the longer term and for the whole basin, which reflects standing wave conditions. Especially near the head, the time lag between maximum flood and high water shows a more diffusive result, than the time lag for the ebb and low water. This is attributed to the fact that flow-width averaged values were taken, so that the lag values reflect the impact of the channel-shoal pattern. Results include the complex process of water exchange between flats and channels, which is held responsible for the diffuse results near the head.

Figure 3.9(a) shows that ebb and flood duration are similar at the mouth and that ebb (flood) duration is larger (smaller) for the remaining part of the basin, although ebb and flood duration become similar over the longer term.

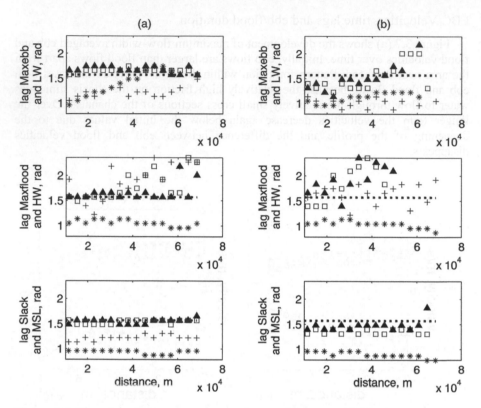

Figure 3.8 *Time lags between maximum flow-width averaged flood velocities, ebb velocities and slack water after high water and (respectively) high water, low water and MSL after high water along the basin for different points in time ('*'=initial condition, '+' = 100 years, '□' = 1600 years, 'Δ' = 3200 years). Data near the head are not shown since this area falls completely dry at relatively low water. (a) For FBC and (b) for EBC. Dashed lines indicate values of ½ π.*

The following section provides a possible physical explanation of the observations. The impact of friction is initially high resulting in a decrease of the tidal amplitude and the tidal velocities along the basin (Figure 3.6 (a) and Figure 3.7 (a) respectively). The relatively small time lag between maximum slack water and MSL [

Figure 3.8(a)] is initially probably caused by friction. Water levels are lower during ebb than during flood. This means that, especially at the intertidal shoals near the head, the impact of friction is high during ebb and the shoals drain their water slowly towards the channels and prolong ebb [Figure 3.9 (a)], whereas inundation takes place more rapidly. As a result of the long ebb duration, ebb velocities are relatively low near the head.

Ebb velocities near the mouth are relatively large compared to the flood velocities. The first reason might be the phase lag between the water levels and velocities, which induces a landward Stokes' drift. Since the embayment is closed at one end, higher mean water levels develop towards the head and, as a consequence, a seaward return flow emerges, which enhances ebb velocities. A second reason is related to higher water level gradients during ebb. For a stationary flow through a uniform cross-section over a horizontal bed, friction leads to a downstream water level gradient. Although this might be somewhat counterintuitive, as a consequence, continuity requires downstream acceleration of flow. In the current models a similar effect might occur, despite the fact that the flow is not stationary and bed levels are not horizontal. The phenomenon will be more pronounced during ebb flow, because of the lower water levels and the higher impact of friction. Figure 3.5 shows that, initially, the lowest water levels in the mouth are lower than the low water level at the relatively deep sea boundary. In contrast, this effect is not observed for the high water levels.

Over the longer term, when the profile deepens, the impact of friction becomes less and more energy of the incoming tidal wave reflects against the head. Resonance occurs and standing wave conditions develop, which results in a time lag of about ½ π [

Figure 3.8(a)], and similar water levels during ebb and flood. The Stokes' drift then disappears, ebb and flood velocities become similar [Figure 3.7(a)], and ebb and flood duration become equal along the basin [Figure 3.9 (a)].

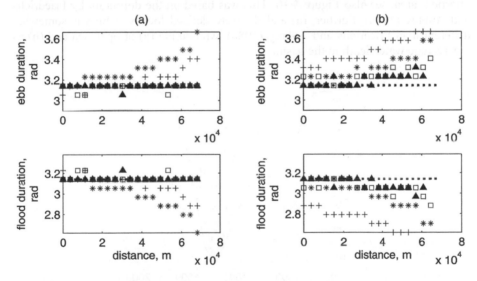

Figure 3.9 *Duration of ebb and flood along the basin for different points in time ('*'=initial condition, '+' = 100 years, '□' = 1600 years, '▲' = 3200 years) based on width averaged values for the velocities and water levels. Data near the head are not shown since this area falls completely dry at relatively low water. (a) For FBC and (b) for EBC. Results are at discrete intervals because they are based on output generated at 10 minute intervals. Dashed lines indicate values of π.*

FBC relation to wave celerity

Based on highly schematized 1D model results, Friedrichs and Aubrey (1988) derived a relationship for morphological equilibrium conditions in short tidal embayments with intertidal area. The relationship is between the ratio of the water volume stored on the shoals and the channel volume (Vs/Vc) and the ratio of the tidal amplitude and the water depth (a/h). Friedrichs and Aubrey (1988) suggest that the duration of flood is increased by water storage on intertidal flats and that ebb duration is prolonged by friction, which increases at lower water levels. Further, they mention the effect of water level on tidal wave celerity, which becomes smaller at relatively low water levels (Airy, 1845) and causes longer ebb duration as well.

Wang et al. (1999) combined the schematized model configuration of Friedrichs and Aubrey (1988) with the assumption of Dronkers (1998) that morphodynamic equilibrium can only be present in case of equal ebb and flood duration. This yields the following equation, which resembles the results of Friedrichs and Aubrey (1988), at least for Vs/Vc < 1, which is the relevant range for the current study:

$$\frac{V_s}{V_c} = \frac{8}{3}\frac{(a/h)^2}{1-a/h}\left(\frac{1+a/h}{1-a/h}\right)\left(\frac{3}{4}+\frac{1}{4}a/h\right)^{-1}$$ (3.1)

For the current research, the shoal volume was calculated by the maximum volume of water present above intertidal area during a tidal cycle. The channel volume was calculated as the volume of water below MSL and not belonging to intertidal area, see also Figure 3.10. This was based on the definition by Friedrichs and Aubrey (1990). Further, (a) and (h) were defined locally, which is somewhat different than Friedrichs and Aubrey (1988), who define (a) at open sea and (h) as the average water depth in the basin.

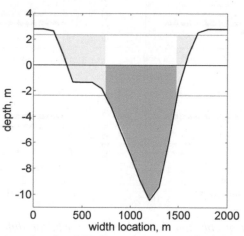

Figure 3.10 *Definition sketch of EBC cross section (thick line) at 50 km from the mouth, including HW and LW (dashed lines) and MSL (thin line). Dark are represents channel volume (Vc). Light shaded area represents shoal volume (Vs).*

Figure 3.11 (a) shows model results of the (Vs/Vc) and (a/h) relationship. The results were obtained by separated model runs on a constant bathymetry (no morphodynamic updating) at different points in time. Comparison of the initial bed level and the bed level after 100 years shows that the development of intertidal shoals has initially a high impact. The deepening of the profile and the increase of amplitude in later stages (> 100 years) lead to a gradual development towards the equilibrium line defined by equation (3.1). The conditions at the mouth move slowly towards the lower left corner. Friedrichs and Aubrey (1988) suggest that the area to the left of the line representing equation (3.1) can be considered ebb dominant, whereas the area to the right is considered flood dominant. Near the head the model results are in the flood dominant domain, whereas towards the mouth they approach the equilibrium condition defined by equation (3.1). Closer analysis of ebb and flood duration along the basin presented in Figure 3.9 (a) leads to similar conclusions. Initially, ebb duration is longer than flood duration, especially in the landward part of the basin. Over the longer term the ebb and flood duration become similar along the basin.

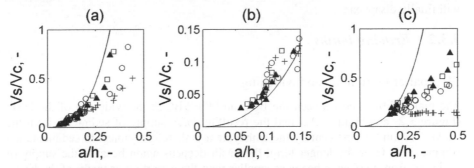

Figure 3.11 *Relation of (a/h) and (Vs/Vc) ratios for model results at different point in time ('+' = 100 years, 'o' = 400 years, '□' = 1600 years, '▲' = 3200 years) and results by equation (3.1) represented by the line. (a) For FBC (b) detail of (a), (c) for EBC*

Friedrichs and Aubrey (1988) suggest that the higher friction during the ebb and that fact that the tidal wave celerity is smaller at relatively low water levels both account for a faster propagating flood wave, a shorter flood period, higher flood velocities and, as a consequence, an importing basin. It is, however, remarkable that the evolution of the current basin is from the flood dominant domain in Figure 3.11 (a) towards a form that reflects morphodynamic equilibrium, which is achieved by a continuously exporting basin [Figure 3.4 (a)]. Apparently, the effect of water storage on intertidal flats and the effects of friction and water depth on wave propagation are not sufficient to explain the evolution of the current model configuration.

A possible explanation can be provided in terms of the Stokes' drift that results from flood flows taking place at higher water levels than ebb flows. In an infinitely long basin the Stokes' drift would thus transport water in the direction of the tidal wave propagation. However, for a closed embayment, mass continuity requires that the drift is counteracted by a seaward directed return flow. This return flow leads to higher ebb velocities [Figure 3.7 (a)] enhancing the ebb dominant and exporting character of the basin.

Based on the presented results one may assume that a form of equilibrium is reached after 3200 years reflecting an equilibrium between the effect of storage (V_s/V_c) and deviations in wave celerity at higher or lower water levels ($c = \sqrt{g(h \pm a)}$). At that point in time, tidal water motion is characterized by a standing wave with maximum ebb and flood velocities taking place at MSL and with approximately similar values and distributions along the basin [Figure 3.7 (a)]. The effect of friction is equal during ebb and flood and may be neglected for its impact on wave propagation for this specific configuration.

Despite the suggested equilibrium conditions, Figure 3.4 (a) shows a continuing sediment export from the basin even after 3200 years. A probable explanation for this is that minor non-linear processes still have an important long-term effect on the basins' morphodynamics. The non-linearities originate from the mass balance equation, both the advection term and the friction term in the momentum equation [Parker et al. (1991)] and the sediment transport formulation. An example of latter is the effect of the interaction between the M2, M4 and M6 tidal constituents on the bed load transport suggested by Van de Kreeke and Robaczewska (1993). It is highly questionable whether the overtides and the associated tide residual transports will finally disappear.

3.3.2   *Eroding banks*

EBC pattern formation and widening

The major difference between the EBC and the FBC is the widening of the basin (Figure 3.2 and Figure 3.12) and the resulting large supply of sediments from the banks. Within the first century the embayment is therefore subject to sedimentation (Figure 3.3). Over the longer term, the basin deepens again because the supply of sediment from the banks becomes smaller than the exporting capacity of the basin. At the same time, a small part of the sediment is transported towards the head where some sedimentation occurs just as in the FBC. Major sedimentation at the head takes place between 400 and 1600 years, which shortens the basin considerably. The exporting process after 3200 years is more pronounced than in the FBC, which is reflected in a higher (gradient of the) accumulated sediment transport after 3200 years (Figure 3.4) and the continuous deepening [Figure 3.3 (b)] and widening [Figure 3.12 (a)] processes.

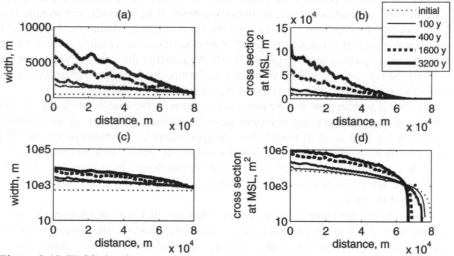

Figure 3.12 *Width development over time along the basin due to bank erosion.*

The widening of the basin appears to be strongly related to channel bends touching and eroding the banks. Because of slow migration of the channel bends and the development of new channels, widening takes place along the complete basin. Initially, there is a pronounced and regular alternating bar pattern present in the basin. After approximately 100 years, especially near the mouth, the main sinusoidal channel grows in amplitude, width and length. Later, after about 1000 years, the main channel breaks up and smaller channels are forming, growing and merging with each other on its remnants. After 3200 years one could distinguish a major channel again having an amplitude of the basin width and a length scale of about half the basin.

Figure 3.2 (b) shows a distorted network of major channels, old channels, new small channels and smaller crosscut channels. The dynamics of the system is still higher than the FBC, and the widening process shows little decline over the long-term. In contrast, Van der Wegen et al. (2007) show that the rate of widening for a 20 km long basin decreases considerably after a period of about 4000 years.

EBC Tidal amplitudes

The initial sedimentation from bank erosion results in an increased damping of the tidal amplitude [Figure 3.6 (b)] due to the influence of increased friction on the tidal hydrodynamics. The water level amplitude then increases again over the longer term. This is attributed to the gradual deepening process, which causes the effect of friction to decrease and enhances resonant behavior, as for the FBC. In addition, the widening process results in a geometry that narrows landward and serves to further amplification of the amplitude

The convergence of estuaries and its impact on the tidal wave characteristics has been subject of extensive earlier research. Friedrichs and Aubrey (1994) suggest that convergence of a basin results in an increase in amplitude. Based on scaling of the shallow water equations they argue that the advection term (the second term in equations I.2 and I.3) can be neglected with respect to the other terms for shallow, strongly convergent estuaries. Consequently, the tidal wave energy through a smaller cross section results in an increase of the water level amplitude. Pillsbury (1956), Jay (1991), Savenije (2001), Savenije and Veling (2005) and Prandle (2003) showed that the water level amplitude can be constant along an estuary with an exponentially decaying cross-section. The effect of friction, which will decrease the amplitude, is then counterbalanced by the effect of the landward decaying cross-section. Typical observation in such an estuary is that the time lag between maximum velocities and maximum water levels is ½ π for semi diurnal tidal conditions and that, although the time lag suggests the presence of a standing wave, the tidal wave is progressive in the sense that it still propagates landward. Pillsbury (1956) referred to this estuary as an 'ideal' estuary, where both the velocity amplitude and the water level amplitude remain constant along the basin. This would lead to constant shear stresses and constant sediment transports.

EBC Velocities, time lags and ebb/flood duration

Model results show indeed that the velocity is relatively constant along the basin for the EBC [Figure 3.7 (b)] and that the time lags between slack water and MSL (observed after high water) tend towards ½ π [

Figure 3.8 (b)]. However, Figure 3.12 shows a linear rather than exponential convergence both in width and cross sectional area along the basin, although one may argue that the cross sectional area has a slightly exponential shape, especially in the area mid-basin towards the mouth [Figure 3.12 (d)]. We suggest three explanations for this. Firstly, a combination of the linear bank convergence and the reflection of the tidal wave against the head causes the ½ π lag in the model results. Research by Friedrichs and Aubrey (1994), Prandle (2003) and Savenije and Veling (2005) do not include reflection of tidal energy against the head as in the current research. Therefore, convergence may remain less than exponential. Secondly, contrary to an 'ideal' estuary, the tidal amplitude [Figure 3.6 (b)] increases along the basin, especially on the longer run. This means that, related to a case with constant tidal amplitude along the basin, relatively large cross sections are needed in the landward part to convey the tidal prism. Finally, contrary to assumptions by Friedrichs and Aubrey (1994), Prandle (2003) and Savenije and Veling (2005) the tidal amplitude becomes large compared to the water depth and final amplitude effects cannot be neglected anymore. While it seems likely that this eventual final amplitude effect will have some influence on the final morphology, it is not straightforwardly explained.

It is expected that model runs in a (much) longer basin and in which the tidal wave completely damps due to friction, would show a more pronounced landward exponentially decaying cross section. In that case the ½ π time lag between water levels and velocities, which is necessary for zero sediment transport gradients, can only develop due to exponential convergence. Finally, a different formulation of the bank erosion algorithm (especially when it leads to a different characteristic timescale compared to the bed level changes) may result in a different width to depth ratio of the basin.

EBC Relation to wave celerity

With respect to the relation between a/h and Vs/Vc, shown in Figure 3.11 (b), the embayment maintains larger values for both ratios than in case of the FBC. Also the basin needs more time to adapt to the profile of equation (3.1). The same observation holds for the time needed to develop towards equal ebb and flood duration, which is shown in Figure 3.9 (b).

It can be concluded that, over the longer term, the EBC responds in a similar manner as the FBC, although the large sediment supply from the banks maintains more morphodynamic action after 3200 years than in the case of the FBC. It is noted that the algorithm for the bank erosion is a rough approximation and does not take explicitly into account the dependency on local values of shear stresses at the bank, sediment/soil characteristics and the exact migration of the banks over time.

3.3.3 Energy dissipation

Taking the expression for energy expenditure of Rodriguez-Iturbe (1992) as a starting point, the work done by the shear stresses and the transport of sediment in a channel section of length L can be described by the following equation

$$ P = c_f \rho_w \frac{U^2}{R} QL + (\rho_s - \rho_w) gSL \qquad (3.2) $$

where

P	energy dissipation, kg m^2/s^3
ρ_w	water density, kg/m^3
U	velocity, m/s
R	hydraulic radius, m
Q	discharge, m^3/s
ρ_s	sediment density, kg/m^3

In terms of the current 2D model parameters and its rectangular grid definition this means that the energy dissipation per grid cell per second at a certain point in time is given by

$$P_{cell} = \left[g \frac{n^2}{\sqrt[3]{h}} \rho_w \left(\overline{u}^2 + \overline{v}^2 \right)^{1.5} + \left(\rho_s - \rho_w \right) g \left(S_x^2 + S_y^2 \right)^{0.5} \right] dxdy \quad (3.3)$$

where

P_{cell}　　energy dissipation in grid cell, kg m^2/s^3

$\overline{u}$　　depth averaged velocity in x direction, m/s

$\overline{v}$　　depth averaged velocity in y direction, m/s

S_x, S_y　sediment transport in x- and y-direction, m^3/m/s

dx, dy　length and width of grid cell, m

In this expression the hydraulic radius (R) was approximated by the water depth (h) under the assumption that the length scale of a grid cell (~100m) is much larger than the water depth (~10m). Results are presented in Figure 3.13 (a) and (c) show the energy dissipation for the total basin over one tidal cycle for the FBC and EBC respectively. The part due to friction increasingly dominates the part due to transport in both cases. A closer analysis of the dominant flow parameters in the energy dissipation process follows from a combination of equations (2.1) (the Engelund-Hansen transport formulation) and (3.3), and leads to the following expression:

$$P_{cell} = \left[\left| \rho_w g n^2 \frac{U^3}{\sqrt[3]{h}} \right| + \left| \frac{0.05}{\Delta D_{50}} \rho_w \sqrt{g} n^3 \frac{U^5}{\sqrt{h}} \right| \right] dxdy \quad (3.4)$$

This shows that, for the current model, the friction term (first term on the right hand side) is less influenced by the water depth and the velocity magnitude than the sediment transport term. Lower velocities and larger depths will thus lead to a more dominant impact of friction on the energy dissipation process. Of course this observation strongly depends on formulation of the sediment transport.

The energy dissipation is initially about 5 times larger for the FBC, since the basin is 5 times wider. The energy dissipation of the FBC increases during the first century due to the pattern formation, resulting in increased velocities and the associated larger shear stresses and sediment transports (Figure 3.7). After one century, the energy dissipation decreases again because of the deeper longitudinal profile and the associated lower velocities. For the EBC, the energy dissipation increases almost continuously due to the widening of the basin, allowing a larger tidal prism to enter the basin [Figure 3.13 (c)].

Figure 3.13 (b) and (d) show the energy dissipation over one tidal cycle averaged over the basin area at MSL for the FBC and EBC respectively. Initially they show similar values. The FBC shows the same trend as Figure 3.13 (a). However, the dissipation in the EBC drops rapidly within the first 50 years Figure 3.13 (e). This is attributed to the high sedimentation of the basin, which leads to shallow depths and low velocities. After 50 years the averaged energy dissipation increases for the next 1600 years. The trend is largest in the first 100 years due to the pattern formation and then the rate of increase reduces. The continuing increasing trend is attributed to sediment supply due to bank erosion, which leads to a relatively shallow basin with a relatively high impact of the channel-shoal system on the behavior of the flow. The flow is more directed through channels resulting in relatively high velocities and energy dissipation compared to the FBC in which the tide propagates on a more submerged bathymetry. As the profile deepens this effect becomes less pronounced and the energy dissipation decreases for the EBC, in a manner similar to the FBC. Remarkably, the value of the energy dissipation after 3200 years is of the same order of magnitude for both cases.

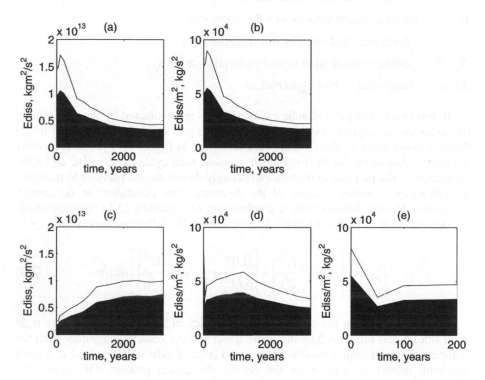

Figure 3.13 *(a) Energy dissipation integrated over the basin area and over one tidal cycle for FBC; (b) Energy dissipation integrated over one tidal cycle per m² for FBC; (c) Energy dissipation integrated over the basin area and over one tidal cycle for EBC; (d) Energy dissipation integrated over one tidal cycle per m² for EBC; (e) detail of (d); Black area representing the friction part and the white area representing the transport part.*

Figure 3.14 presents the distribution of energy dissipation over a tidal cycle and shows how this varies over the period modeled (3200 years). The Figure 3.shows that initially, for both basins, the energy dissipation during ebb exceeds the dissipation during flood. However, after approximately 3200 years the magnitude of the energy dissipation during ebb and flood is almost similar. Further, there is a shift in timing of the dissipation peaks related to the increasing time lag between maximum water levels and maximum velocities (see Figure 3.8).

Figure 3.15 presents results of the tide averaged and flow width averaged energy dissipation along the basin. This shows that the energy dissipation is, in the end, not only decreasing with time, but it is also more uniformly distributed along the basin. The pronounced peaks in the dissipation for the FBC after 3200 years are associated with the presence of bars. This is shown in more detail in Figure 3.16 (b), which represents the energy dissipation in the basin in relation to depth contours. The energy dissipation peaks for a morphodynamically relatively stable situation [FBC after 3200 years, Figure 3.16 (b)] are located at the shoals, whereas more dynamic systems [FBC after 400 years, Figure 3.16 (a) and EBC after 3200 years, Figure 3.16 (c)] show major energy dissipation taking place in the channels. This can be explained by the fact that the initial pattern develops on a bathymetry that is on average too shallow. Largest velocities and major transports and shear stresses occur in the channels. As time goes by, the bathymetry deepens and velocities (and the associated shear stresses and transports) decrease. The deepening may continue to such an extent that the shoals remain under water during the tide. Majority of the shear stresses then takes place at the crests of the flooded shoals. As an example, the seaward shoals in Figure 3.16 (b) are completely drowned (with energy dissipation peaks at the crest), whereas towards the head the shoals are present as intertidal area (with energy dissipation peaks in the channels).

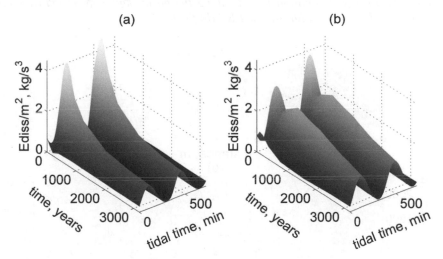

Figure 3.14 *Energy dissipation per m^2 over time (in years) during one tidal cycle. The first peak is related to flood and the second peak to ebb; (a) for FBC (b) for EBC.*

It can be argued that the energy dissipation peaks at the shoals are a consequence of the confinement of the FBC, because it forces the flow to 'bounce' over the shoals. The EBC would not lead to such a system since, during its morphological evolution, the major flow would continuously use channels. This is reflected for the EBC in Figure 3.15(b) showing no pronounced peaks to the extent present in the FBC [Figure 3.15(a)].

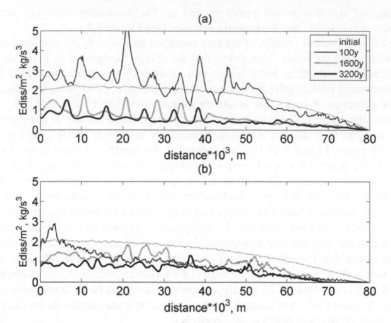

Figure 3.15 *Energy dissipation per m^2 along the basin (kg/s^3) averaged over one tidal cycle and the flow width; (a) for FBC (b) for EBC.*

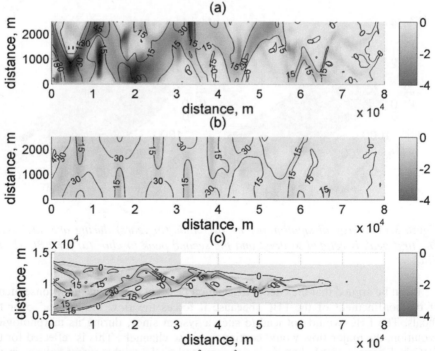

Figure 3.16 *Energy dissipation per m^2 (kg/s^3) averaged over one tidal cycle and depth contour lines with respect to MSL. Negative values indicate higher dissipation; (a) FBC 400 years (b) FBC 3200 years (c) EBC 3200 years.*

3.3.4 Validation

Data on morphological evolution in estuaries and the assumed decreasing energy dissipation over relatively long time spans (~ centuries) are not readily available, because detailed measurements have only been carried out during the last decades or century. Furthermore, where relatively long-term data are available, human intervention (i.e. by access channel dredging, the construction of flow mitigating structures or land reclamation) has invariably had a high impact on the estuarine bathymetry. Finally, sea level rise would also be of importance on a time scale of centuries.

When values of energy dissipation are derived using a 2D numerical model for a specific estuary, based on different historical bathymetries, this leaves the question open as to whether this should be attributed to the system, human intervention or sea level rise. For example, research by Lane (2004) and Blott et al. (2006) shows major changes of the Mersey estuary bathymetry (UK) over the last century, which is attributed to training wall construction and dredging. Their research was based on analysis of historical bathymetric charts and runs of a numerical flow model. However, no distinction could be made between the subsequent effects of sea level rise and eventual wave climate changes or autonomous evolution (the latter is not mentioned in their research).

Validation may take place in terms of the order of magnitude of the actual energy dissipation for a numerical model based on a real geometry and bathymetry. Winterwerp et al. (2001) show an application of such a model considering the Western Scheldt, whereas Kuijper et al. (2004) give details on the validation of a similar Western Scheldt model, in which geometry, grid cell sizes, time steps and forcing were similar to the current model. Figure 3.17 shows an overview of the results of this model based on a 1998 bathymetry and applying astronomic boundary conditions.

The Western Scheldt itself is a relatively young estuary that developed early in the Middle Ages when a tidal channel penetrated landward towards the Scheldt River, located north of Antwerp, where it initially developed over a bed of marshland [Van der Spek, 1997 and Beets and Van der Spek (2000)]. Alluvial sand and remnants of erosion resistant peat layers determine the present bathymetry. Land reclamation taking place over several centuries resulted in dikes constructed on the banks of the Western Scheldt, including scour protection at the bed sometimes extending into the deepest channels adjacent to the dikes.

Figure 3.17 shows that the order of magnitude of energy dissipation is similar to the model results. The peak dissipation found in the relatively narrow section at km 40 from the mouth and the smaller peaks along the basin are attributed to narrowing by land reclamation works and non-erodible peat layers. This can be considered as a physical constraint of the system, preventing it from evolving towards a most probable state with a more gradual and uniform distribution of energy dissipation, which will be subject of discussion in the next section.

Figure 3.17 shows only part of the Western Scheldt which in reality extends more landward and narrows extensively into the Scheldt River. This is outside the domain of the current model, which focuses on the relatively broad and tide dominated area of the Western Scheldt. The presence of the river Scheldt may explain the relatively constant energy dissipation value still present at 60 km from the mouth, contrary to the zero energy dissipation of the current model, and suggests that still considerable tidal energy is present in this section. This is confirmed by

Van der Spek (1997) and Kuijpers et al. (2004) who describe large landward amplifications of the tidal amplitude that only damp beyond Antwerp.

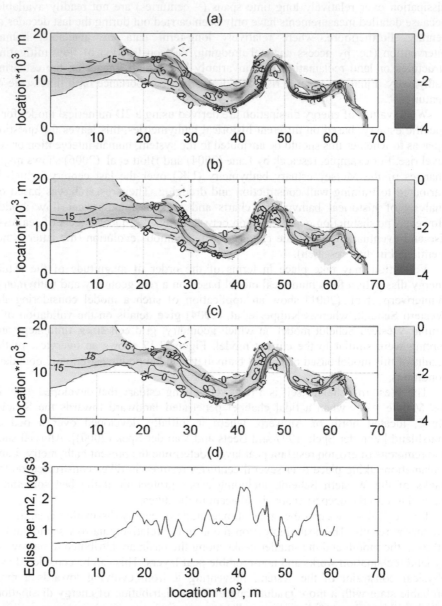

Figure 3.17 *Western Scheldt energy dissipation per m² (kg/s³) averaged over one tidal cycle and depth contour lines with respect to MSL. More negative values indicate higher dissipation. (a) Total energy dissipation; (b) contribution of friction (c) contribution of sediment transport; (d) Energy dissipation per m² along the basin (kg/s³) averaged over one tidal cycle and the width. Cross-sections were taken perpendicular to the main flow direction*

3.4 Discussion

3.4.1 Grid resolution

The results presented thus far, assume that the channel shoal pattern may be described by grid cell sizes of 100 m by 200 m. This puts a limit to the minimum spatial scale the model is able to resolve. On the seaward side of the embayment this may be acceptable, since in that area large tidal prisms generate relatively large spatial patterns. For the cross section, for example, this is confirmed by empirical relationships between tidal prism and cross sectional area suggested by Jarrett (1976). Further, Van der Wegen and Roelvink (2008) suggest a relationship between a characteristic longitudinal morphological length scale and the tidal prism based on their model results. Also the characteristic morphological length scales are much larger than the applied grid size. More landward, however, the tidal prism and channel cross section become smaller and grid resolution starts to have a larger impact on an accurate description of the channel cross section, up to the extent that the grid cannot describe the channels anymore.

An additional model run was carried out with a grid size of 33m by 67m, which is three times smaller than in the original model. Because of the long calculation duration involved (~ 4 weeks), this was done for a relatively short, 20 km long, basin for 600 years with similar model settings as the current model [see for a closer description of the short basin model with coarse grid Van der Wegen et al. (2007)]. The geometry of this model resembles the landward final 20 km of the current model, in the sense that it has equal basin size and that the bed levels and percentage of intertidal area are similar.

Comparison of Figure 3.18 (a) and (b) shows that major pattern characteristics are maintained for the finer grid, especially near the mouth, where one distinguishes a single main channel with similar size. Compared to the coarse grid model, the fine grid model shows more detail especially near the head (i.e. shoals within the main channel, more channel curvature and the presence of smaller channels) and the difference with the coarse grid model in that area is relatively large. The comparison suggests that seaward the impact of grid size reduces and that small scale patterns do not impact on the major morphodynamic features. No distinct differences were observed in the width averaged longitudinal bed profiles [Figure 3.18 (c)]. For the current 80 km long basin, this suggests that the coarse grid qualitatively resolves major pattern characteristics for a large part of the long basin.

3.4.2 Sea level rise

Fleming et al. (1998) conclude that, after a period of rapid sea level rise, sea level has been relatively constant over the past 6000 years, although still uncertainty exists on the extent of fluctuations during this period. Based on observations, Dronkers (1986), Woodroffe et al (1993), Pethick (1994), Rees et al. (2000) and Beets and Van der Spek (2000) suggest that the relatively constant sea level created conditions for considerable infill of different basins and river valleys that were previously drowned by the rapid sea level rise.

Sea level rise is absent in the current model set-up. In this way, development of equilibrium could be investigated under constant boundary conditions. The current research suggests that major morphodynamic adaptations of alluvial coastal plain estuaries are covered within the timeframe of the recent relatively constant sea level (~millennia), provided that there is enough sediment available. Concerning the small life span of about 800 years of the Western Scheldt estuary and the fact that it developed over shallow, drowned marshland, model results would imply that the estuary is yet far from equilibrium. However, for a more thorough analysis a number of additional processes should be taken into account, like, for example, bed composition, the existence of non-erodible layers, wave driven transport at the mouth and the bathymetry seaward from the mouth. Furthermore, this research leaves the question how coastal plain estuaries would respond to the relatively small fluctuations in sea level rise over the past 6000 years suggested by Fleming et al. (1998). These small fluctuations in sea level, not only affecting MSL but also the character of the tide at the mouth, could have a considerable impact. This will be subject of future research.

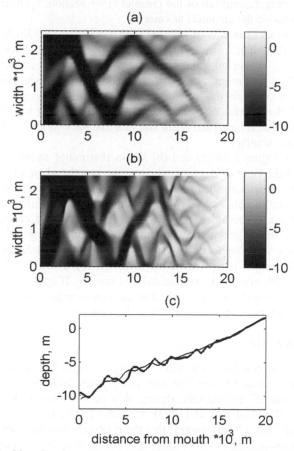

Figure 3.18 *Bed levels after 600 years for 2.5 km wide and 20 km long basin with (a) coarse grid 100m x 200m (b) fine grid 33m x 67m and (d) width averaged longitudinal bed profile for coarse grid (dashed line) and fine grid (line).*

3.4.3 Initial conditions

Contrary to the observed basin infill due to sea level rise mentioned in the last section, the current model shows a continuous sediment export. One explanation for this might be the modeling approach not taking into account finer sediments and settling lag effects. Another possible explanation is that the initial condition was taken too shallow in order to stimulate pattern formation and to relate it to the current Western Scheldt bathymetry.

A number of reasons may be responsible for the shallowness of the Western Scheldt estuary. Firstly, it developed only about 800 years ago, which means that it was not drowned to the extent of river valleys during the period of rapid sea level rise and that the period for pronounced further deepening has been too short. Secondly, the presence of non-erodible layers may have prevented the Western Scheldt basin from deepening further. However, Dronkers (1998) considers these reasons not being dominant suggesting that the estuary is close to equilibrium. He attributes this is to a strong M4 component being present outside the mouth on the relatively shallow continental shelf. The current model only prescribes a sine shaped semi-diurnal water level at the seaward boundary and would therefore not lead to importing conditions. Further, wave induced transport from adjacent coastlines and the ebb tidal delta, the occurrence of storm events and sea level rise might play an important role as well.

It is questioned whether another, for example deeper or more gently sloped, initial condition would lead to similar results. Sensitivity analysis in this respect was not carried out, because of the long computation time involved (~ months). Another problem would be the long adaptation timescale that may finally lead to a similar result, but not within the timeframe considered (~4000 years). This would especially hold for configurations with deeper initial conditions which will transport less sediment. It is believed that similar results will be obtained in terms of characteristic morphological length scales, especially in relation to local flow conditions. The width averaged longitudinal profile could be subject to (apart from the local disturbances by shoals) a horizontal shift landward when the initial profile is relatively deep and seaward for shallower initial conditions [for a 1D configuration this is confirmed by Van der Wegen and Roelvink (2008)]. The seaward shift will be caused by rapid sedimentation at the head. Since the impact of friction for such a shallow basin remains relatively high, the tidal amplitude may remain damped and no resonant tidal behavior may occur.

Another important initial condition is that the banks were set at a level higher than the expected highest water level in the basin. This was done since preliminary calculations showed considerable infill of the basin when banks were flooded at a certain moment due to the resonant tidal wave. This delayed the evolution considerably and complicated the analysis of the results. Furthermore, lower bank levels (below high water or even below MSL) would lead to large and extending amount of intertidal area and a system quite different than the reference case of the Western Scheldt estuary. Marciano et al. (2005) describe an example of the evolution of such a basin, i.e. a typical inlet in the Dutch Waddenzee area, following a similar process-based approach. A more extended intertidal area would probably influence final basin characteristics by means of different intertidal flat flooding and draining characteristics (described, for example, in more detail by D'Alpaos et al. (2005) for the Venice lagoon), and its impact on tidal wave propagation and dominant sedimentation processes. These may include for example settling lags and erosion lags and the impact of vegetation as described by D'Alpaos et al. (2007) and Murray et al. (2007). Whether or not such a system would finally lead to decreasing basin wide energy dissipation needs to be confirmed by future research.

Considering further the impact of initial conditions on energy dissipation, for optimal fluvial networks, Rinaldo et al (2006) suggest that initial (and boundary) conditions affect the feasible optimal state of the network. This means that initial conditions influence the final values of minimum energy dissipation so that different initial conditions could possibly lead to a more optimal network state as well as lower values for the energy dissipation. This may also hold for this tidal configuration especially considering the assumed shift of the longitudinal profile. Closer analysis is considered to be outside the scope of the current research.

3.4.4 Energy dissipation

The major processes taking place in the basins (short-term pattern formation and long-term deepening and widening) are well reflected by the evolution of the energy dissipation. Also, the lower state of morphodynamic activity, especially apparent in the FBC for the period considered, agrees well with the decreasing (rate of) energy dissipation.

Rodriguez-Iturbe et al. (1992) use their energy dissipation formulation to show that their considered drainage network develops towards a state of minimum energy dissipation. The current model leads to a similar conclusion. However, Rodriguez-Iturbe et al. (1992) consider a loop less network with fixed links between flow sections, whereas the current research describes a dynamic evolution, that allows for slowly migrating bars and channel configurations, local peaks in energy dissipation shifting along the basin, flow across tidal flats, and even loops developing in the channel network. Therefore, it is difficult to compare the current model results to their first two underlying principles (i.e. the principle of minimum energy dissipation in any link of the network, and the principle of equal energy dissipation per unit channel anywhere in the network), although their third and final principle of minimization of energy dissipation in the network as a whole seems to be confirmed by the current research considering the current model domain as the "network".

More pronounced and in apparent contradiction to these three principles are the effects of pattern formation and bank erosion, since both processes lead to a temporal increase in the overall basin energy dissipation. This is attributed to the initial condition prescribing a flat bed, which imposes an inherent instability to the system that the processes seek to adjust. During a representative adaptation period, in which the dominant patterns develop, the geometry is formed that fulfils the conditions of the system. After the adaptation period the energy dissipation decreases at a decreasingly lower rate and even becomes less than at the initial, flat state.

3.4.5 Energy flux

These findings can be linked to equilibrium conditions derived from an entropy based approach in research by Leopold and Langbein (1962) and Townend (1999) and Townend and Dun (2000). The latter two references are more specifically related to estuarine environments and use the criterion of constant entropy production rate per unit discharge to suggest that the longitudinal energy flux distribution along an estuary can be described by an exponentially decreasing function. Their reasoning is briefly repeated here for reasons of clarity.

Leopold and Langbein (1962) introduce the principle that river systems aim for least work, which may be interpreted as minimum production of entropy. They suggest that this state will be reached when the rate of entropy production per unit discharge is constant along a river and is proportional to the longitudinal gradient of energy dissipation, which is expressed in the following relation:

$$\frac{1}{Q}\frac{\mathrm{d}\varphi}{\mathrm{d}t} = \frac{1}{H}\frac{\mathrm{d}H}{\mathrm{d}x} = a \qquad (3.5)$$

Where

$d\varphi$	entropy production, m^2
dt	time span, s
dH	loss of energy head over distance dx, m
Q	discharge, m^3/s
H	energy head relative to some datum, m
a	constant, m^{-1}

Langbein (1963) applied this theory to estuaries as well, albeit for a highly schematized case in which the amplitudes of the water level and the velocity are assumed constant along the estuary and in which friction and overtides are disregarded.

Considering that estuaries, contrary to river systems, are governed by time and directionally varying discharges over the tide, Townend (1999) suggests that the rate of entropy production per unit discharge for estuaries can be expressed by evaluating the energy flux instead of the energy head. He thus proposed:

$$\frac{1}{Q}\frac{\mathrm{d}\varphi}{\mathrm{d}t} = \frac{1}{HQ}\frac{\mathrm{d}(HQ)}{\mathrm{d}x} \qquad (3.6)$$

Further, Townend (1999) follows the suggestion by Leopold and Langbein (1962) that the most probable state of the estuarine geometry would be reached when the rate of entropy production is constant. This yields:

$$\frac{1}{HQ}\frac{\mathrm{d}(HQ)}{\mathrm{d}x} = a \qquad (3.7)$$

Solving equation (3.7) and considering integration over a complete tidal period (T) yields with (α) and (β) being constants:

$$\rho g \int_0^T HQ \mathrm{d}t = \exp(\alpha x + \beta) \qquad (3.8)$$

Equation (3.8) suggests that for a constant production of entropy per unit discharge the energy transported due to the tidal wave will decay exponentially landward. Townend and Dun (2000) showed that the equation leads to good results for the Humber estuary and the Bristol Channel in the UK. In case of the Humber small discrepancies between equation (3.8) and 1D model results were attributed to the effect of river discharge and the effect of the construction of training work. Even evolution towards a better fit to equation (3.8) was suggested by using different historical bathymetries ranging over 150 years. For the Bristol Channel deviations between equation (3.8) and a 2D model were attributed to the geology of the basin having only limited sediment supply and hard banks, both opposing further evolution towards the most probable state. Comparison to a third estuary, Southampton Water, showed less favorable results. 1D model results showed a linear relationship between distance from the mouth and energy flux. This was attributed to the rather rectangular geometry of the basin with only little supply of sediments for further evolution.

Figure 3.19 shows the evolution of the current model in terms of the left side of equation (3.8) taking absolute values for (HQ). Apart from the anomalies caused by the channel-shoal pattern, the FBC continuously shows a linear relationship between distance from the mouth and energy flux just like in case of Southampton Water. The linear relationship suggests uniform work or constant energy dissipation along the basin, contrary to equation (3.7):

$$\frac{d(HQ)}{dx} = \text{constant} \tag{3.9}$$

Occurrence of uniform energy dissipation is supported by Figure 3.15 (a).

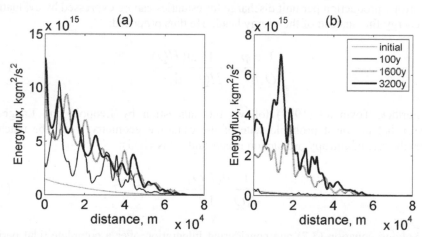

Figure 3.19 *Width integrated energy flux distribution along the basin. (a) FBC and (b) EBC.*

Compared to the FBC, the EBC shows a more exponential relationship similar to equation (3.8), although it only becomes apparent on the longer term and, still, longitudinal deviations are large due to the channel-shoal pattern. Also, the exponential distribution is not outspoken and may even be questionable. A linear distribution is more supported by the fact that, like in the fixed basin configuration, longitudinal energy dissipation is rather uniform along the basin [see Figure 3.16(b)] and by the fact that the basin shows a linear rather than an exponential widening of the banks towards the mouth, see Figure 3.12.

The fact that both configurations aim for uniform distribution of energy dissipation along the basin, contrary to uniform work described by equation (3.7), suggests that the most probable state of the system described by the shallow water equations and the Engelund Hansen transport formulation is not reached. Assuming the validity of equation (3.7), a first possible explanation could be that the linear decrease of energy head observed along the estuary does not necessarily contradict equation (3.7). This holds, for example, for the case when the discharge integrated over the tide decreases exponentially along the basin, whereas the energy head still decreases linearly. Equation (3.7) is then dominated by the along basin discharge profile.

A second possible explanation is the occurrence of inherent constraints in the definition of the model configurations. Apart from the definition of fixed banks, the prescribed length of the basin can be considered a constraint as well. On the longer run, resonance develops in the basin leading to larger water level amplitudes, especially near the head. For the EBC, the defined bank level at MSL +3.2 m prevents the areas near the head from flooding. In case that the basin would not have been blocked at the head and that bank level would have been lower, flooding would occur at the head due to the landward increasing tidal amplitude. This would possibly result in a longer basin with a landward exponentially converging width. Such a basin could lead to better agreement with equations (3.7) and (3.8). The tidal dynamics of such a long, strongly converging type of estuary have been extensively studied [amongst others by Jay (1991), Friedrichs and Aubrey (1996), Prandle (2003), Savenije and Veling (2005) and Toffolon et al. (2006)] although none of the studies were carried out based on an explicit morphodynamic formulation like in the current process-based approach.

The Western Scheldt estuary has weakly converging width [Savenije and Veling (2005) and Toffolon and Crosato (2007)]. Despite the fact that the size of the current model configuration was inspired by this estuary, the model does not reproduce such geometry. Assuming that the current models' bank erosion algorithm is adequate enough, probable reasons for this are the facts that the Western Scheldt is relatively young, its banks are protected by dikes and that its 70 km long narrow river section is not taken into account in the current model.

3.5 Conclusions

The long-term, process-based approach of the current research shows evolution of morphodynamic features and timescales that are comparable to estuarine morphodynamics found in reality. This is despite a relatively simple system of shallow water equations and a straightforward formulation of the sediment transport. A major shortcoming is the computational time required (~ weeks for 3200 years), so that relatively few runs have been carried out for the current research.

The basins under consideration evolve towards a state of less morphodynamic activity, represented by relatively stable characteristic patterns and decreasing deepening and widening of the basins. This is despite observations that there remains a continuous sediment export towards the sea and that bar and shoal migration continues to take place. Closer analysis of the hydrodynamic characteristics over time shows lower width averaged velocity amplitudes, similar ebb and flood duration and velocities along the basin, constant time lags between maximum velocities and maximum water levels and a stable (a/h) and (Vs/Vc) relationship. With respect to the latter, it is suggested that equilibrium conditions are formed by two balancing effects on the tidal wave celerity, namely the effect of storage and the effect of tidally fluctuating water levels. The effect of friction on the tidal asymmetry, implicitly also present in the factor (a/h), can apparently be neglected. A probable explanation is that after 3200 years both basins reflect a situation in which a standing wave is present. In that case, ebb and flood velocities take place at equal water levels and their magnitude becomes similar, so that the effect of friction is equal during ebb and flood and it has no effect anymore on tidal asymmetry.

The current research shows that energy dissipation reflects the major processes and their timescales (pattern formation, widening and deepening). Pattern formation and bank erosion both cause an initial and temporal increase in energy dissipation. On the longer term, however, the basin wide energy dissipation decreases at decreasingly lower rate and becomes more uniformly distributed along the basin.

Comparison of the model results to a 2D numerical model describing a comparable basin (Western Scheldt) shows that the order of magnitude of the energy dissipation and its relatively uniform width-averaged longitudinal distribution are indeed similar. Deviations are attributed to bank protection works and the small current model domain that excludes the narrow landward section of the Western Scheldt and in which the tidal wave propagates.

Comparison of the model results to equilibrium conditions defined by minimum entropy production, or, constant entropy production rate, suggests that the fixed banks and the prescribed level of the banks in combination with the prescribed length of the estuary prevent the chosen hydrodynamic and morphodynamic model formulations from evolving toward the most probable state.

4 Morphodynamic modeling of tidal channel evolution in comparison to empirical PA relationship[3]

Abstract

Tidal channels and inlets in alluvial environments are interconnected dynamic systems that react to changing physical conditions (such as sea level rise) as well as to anthropogenic impact (such as dredging and bank protection works). Past research resulted in an empirical equilibrium relationship for inlets between the tidal prism (P) and the cross-sectional area in a tidal inlet (A) and constant PA relationships were found along several tidal basins.

Physical explanations of the PA relationship are based on a balance between littoral drift and tidal transport capacity (for inlets), on the concept of critical shear stress for sediment transport or on the appearance of constant spatial sediment concentrations, which is related to spatially uniform shear stresses. The current research addresses the PA relationship by investigating the morphodynamic evolution of a tidal embayment over millennia. Use is made of a 2D process-based, numerical model that is capable of describing long-term morphodynamic basin evolution and pattern development. Wave impact (i.e. littoral drift) is neglected so that focus is on tidal motion only.

Model results show that a shallow basin mouth attached to a deep ocean and a deep basin continuously deepens, although development rates decrease over time. Basin mouth evolution becomes strongly damped when it is attached to a shallow basin that exports sediment. The PA relationship is constant along the shallow basin which is attributed to constant tidal conditions along the basin. The PA relationship in the mouth and the adjacent tidal embayment compares well with the empirically derived PA relationship by Jarrett (1976). However, on long (~century) time scales model results show slow, but continuously increasing cross-sections for similar tidal prisms. This tidal channel evolution is related to (very) small but persisting spatial gradients in tide residual transport.

Further analysis shows that confined inlets require smaller cross-sectional areas than free inlets to convey the same tidal prism, which is attributed to their relatively deep cross-sectional area so that friction plays a smaller role. This is in line with different coefficients found for jettied inlets and free inlets in the empirical PA relationship

---

[3] A slightly adapted version of this chapter was submitted to Coastal Engineering by authors Van der Wegen, M., A. Dastgheib and J.A.Roelvink.

4.1 Introduction

Tidal channels convey flows in tidal embayments and are often organized in tidal channel networks (Rinaldo et al. 1999a). On the landward side they may emerge from tidal flats, but they can also be connected to river channels. In seaward direction, several tidal channels may merge into larger channels, so that eventually one or two major tidal channels are left which discharge into the ocean via a tidal inlet. In coastal engineering inlets are often referred to as the narrow gorges between ocean and a tidal basin or lagoon. Bruun (1978) applies a wider definition and distinguishes three types of inlet, namely those with a geological origin (determined for example by rocky outcrops), those with a hydrological origin (determined by river outflow) and those with a littoral drift origin (mainly governed by wave induced sediment transports and tidal movement).

Many inlets and channels are used as access channels to ports located in areas sheltered from immediate (wave) impact. In order to maintain a specific depth some inlets or channels are regularly dredged or jetties have been built to protect the inlet from infilling by sediments from littoral drift.

In an alluvial environment, the stability of tidal channels and inlets is related to morphodynamic processes in adjacent morphological units. Apart from the tidal inlet and the channels themselves, examples of these units are the coastline, ebb tidal delta, tidal bars, channel networks and tidal flats and salt marshes in the embayment (De Swart and Zimmerman (2009)). Examples of morphological evolution are seasonal inlet closure, decadal inlet migration, basin reduction by land reclamation works or impacts of sea level rise.

Knowledge on the morphodynamic character of inlets and tidal channels is relevant, since optimal configurations are needed to minimize dredging costs and to prevent unwanted morphodynamic behavior. Another motivation is that adjacent areas (such as protected wetlands) need to be managed as adequately as possible. Therefore, morphodynamic behavior and the associated assumed equilibrium in tidal inlets and channels have been subject of extensive research.

4.1.1 PA relationship

Initially, empirical research focused on the relationship between the tidal prism through an inlet (P) and a characteristic cross-sectional area (A) of the inlet (further referred to as the PA relationship), which can be represented by the general formulation

$$A = CP^n \qquad (4.1)$$

in which
A   characteristic cross-sectional area, m^2
C   constant, unit depending on 'n'
P   tidal prism, m^3
n   coefficient, -

Empirical research

Based on measurements of harbor entrances along the Pacific US coast, LeConte (1905) found n=1, and distinguished between unprotected entrances and entrances with jetties by attributing 30 % higher values for C for protected entrances. O'Brien extended the research to measurements of tidal inlets along the Pacific coast (O'Brien, 1931) and the coasts of the US Gulf and Atlantic (O'Brien, 1969), where 'A' was defined as the cross-section below MSL taken in the narrowest part of the inlet. Further, Jarrett (1976), by integrating data of the former studies and including new measurements, assigned different values for 'C' and 'n' depending on the location in the United States with specific tidal conditions (Pacific, Gulf or Atlantic) and the character of the inlet (i.e. unjettied or with a single jetty and with two jetties).

Based on an analysis of UK estuaries Townend (2005) improved the fit to the O'Brien (1931) relationship by including the ratio of estuarine length and the tidal wavelength. Hume and Herdendorf (1993) verified the PA relationship for 16 estuary inlets in New Zealand with a different type of geological background, ranging from barrier enclosed basins to embayments with a volcanic bathymetry. They concluded that the PA relationship is obeyed, although they distinguished, similar to Townend (2005), different values for C and n for different types of estuaries.

Earliest empirical research on the PA relationship has had a strong focus on the inlet mouth. However, the relationship appears to be valid along tidal channels as well. De Jong and Gerritsen (1984) and Eysink (1990) observed a constant PA relationship along the Western Scheldt estuary, Friedrichs (1995) summarized observations for other estuaries and sheltered embayments and Rinaldo et al. (1999b) and D'Alpaos et al. (2010) found similar results from measurements in the Venice lagoon. This suggests that the PA relationship is valid for a larger spatial scales and a larger range of tidal environments. However, due to the presence of waves and littoral drift, the physics behind the PA relationship in inlets will probably be distinct from more sheltered regions in the basin such as tidal channels.Physical explanations

For tidal inlets attempts have been made to clarify the coefficients of the PA relationship in a more theoretical way. Bruun (1978), Kraus (1998) and Van de Kreeke (1998, 2004) derived expressions based on the balance between long shore, wave induced supply of sediments to the inlet and the mainly tidally induced (ebb)flow capacity to transport this sediment from the inlet. Van de Kreeke (2004) and Walton (2004) additionally included the impact of cross-sectional dimensions. These physical explanations are related to inlets subject to littoral drift. An exception is the expression derived by Van de Kreeke (2004) allowing for absence of littoral drift, but this leads to an infinitely large cross-sectional area.

Other attempts to physically explain the PA relationship are based on the concept of critical shear stress. These explanations assume local inlet equilibrium based on sediment properties and hydrodynamic conditions. Krishnamurthy (1977) derived an expression for a minimum required cross-sectional area based on a critical shear stress for sediment motion and the assumption of a logarithmic velocity profile. Hughes (2002) extended the work by including the impact of different cross-sectional areas. The expression by Hughes (2002) still includes a dimensional coefficient that contains, amongst others, the effects of non-sinusoidal tides. D'Alpaos et al. (2009) and references therein and Tambroni and Seminara

(2010) again explain the PA relationship assuming the critical shear stress concept, although Tambroni and Seminara (2010) point to the possible presence of a dynamic equilibrium in inlets subject to littoral drift in contrast to an assumed static equilibrium in adjacent tidal channels. Both Krishnamurthy (1977) and Hughes (2002) focused on inlet stability although their explanation could be valid for more sheltered areas as well. Friedrichs (1995) suggested that, for embayments sheltered from waves, the critical shear stress criterion would lead only to an upper boundary for equilibrium cross-sectional area along the embayment. Friedrichs (1995) thus implicitly suggests that processes other than littoral drift play a role as well. For example, Friedrichs (1995) and Gao and Collins (1994) point to the importance of zero spatial gradients in tide residual sediment transports along a basin or inlet for maintaining equilibrium. Further, Gao and Collins (1994) addressed the possible impact of fresh water discharge on final equilibrium..

Finally, physical explanations of the PA relationship are described based on the concept of equilibrium sediment concentration. Gerritsen et al (2003) derived an expression based on the assumption suggested by Di Silvio (1989) that an equilibrium inlet cross-section requires an equilibrium concentration of sediment in the inlet and combined this assumption with the Engelund-Hansen transport formulation (that does not include a threshold value for sediment motion). Still Gerritsen et al (2003) used empirically derived dimensional coefficients in their final equilibrium expression.

Interaction between inlet and other morphological units

Inlets are morphologically related to adjacent morphological units such as the coastline, the ebb delta in the ocean or the flood delta in the basin and tidal channels in the basin itself. An example of an aggregate modeling approach that is based on the interaction (of some of) these units, is given by Stive et al. (1998), Stive and Wang (2003) and Van de Kreeke (2004, 2006). Key factor in the approach is that it prescribes and applies empirical relationships in the morphological units. Distortion of any of these morphological units leads to morphodynamic interaction and adaptation of the whole system until equilibrium is restored. For example, Van de Kreeke (2004, 2006) considers the evolution of an inlet and its basin in the Dutch Waddenzee and estimates the reaction timescale of an inlet by basin adaptations. Although good results were obtained on inlet evolution on a decadal timescale, the method just applies an empirically derived equilibrium without physically explaining it. It does not include and predict, for example, possible slow modifications of the PA relationship over time.

4.1.2   Aim and approach

Literature on inlet and tidal channel morphodynamics recognizes dominant parameters and processes. Theoretical models lead to acceptable correspondence with empirically derived relationships. However, tidal inlet research focused on equilibrium between tidal sediment and (considerable) wave induced littoral drift. Other possible sources of sediment supply towards the inlet, such as the adjacent basin, have been merely disregarded. Less attention has been paid to equilibrium in the absence of waves and for conditions in the basin which are sheltered from wave impact. Also, long-term inlet and tidal channel evolution has not been studied in detail.

The aim of the current research is to investigate the physical processes underlying the PA relationship with special emphasis on the interaction between inlet and basin. Use is made of a 2D process-based, numerical model that is able to describe time evolution and channel–shoal pattern development.

Central hypothesis is that, in the absence of waves, the PA relationship along the basin and in the basin mouth is comparable with the empirically derived relationship and that it is constant at a particular moment in time, but that equilibrium is not present on longer time scales (~centuries). Since a tidal inlet is associated with the impact of waves and littoral drift, in the current research (disregarding waves) the region between basin and ocean is further referred to as basin mouth.

First we will investigate the development of a shallow basin mouth situated between a deep sea and a deep basin. Initial focus is on a (too) small basin mouth cross-section to study the time evolution of the basin mouth in terms of its PA relationship. Secondly, the basin is made shallow so that it acts as sediment source to the basin mouth. Both the basin mouth development and the development within the basin will be evaluated in terms of the PA relationship over a timeframe of millennia. A detailed analysis is made on the impact of changing conditions like sediment supply from the basin, basin geometry and tidal wave characteristics

4.2 Model set-up

The current research applies a 2D, process-based, numerical model (Delft 3D). For a particular configuration the model solves the 2D shallow water equations, calculates the sediment transport and updates the bed according to the divergence of the sediment transport field. The model includes bed slope and bank erosion effects. More details on the numerical aspects can be found in Lesser et al. (2004). An advanced morphodynamic update scheme that adapts the bed every time step (Roelvink (2006)) was used in order to speed up morphodynamic developments. Further, Van der Wegen et al. (2007, 2008) describe detailed model settings, similar to the current research and include the Engelund-Hansen transport formulation and a formulation of wetting and drying procedures.

The model configuration consists of a narrow 2.5 km wide basin that is connected to a deep sea. The basin mouth is defined as the short section where the basin reaches the sea, see Figure 4.1a. The system is forced by a semi-diurnal tide with water level amplitude of 1.75 m that approaches the coast perpendicularly. The bed level at sea is set at 30m-MSL so that friction effects will not be significant in this area. Different initial bathymetry and geometry settings are applied. The first setting describes a deep basin and a shallow basin mouth section, so that basin mouth development without sediment supply from the basin side can be investigated; see Figure 4.1b for a 20 km long basin geometry. In order to study the effect of sediment supply from the basin, the second setting describes a deep sea and a shallow basin, see Figure 4.1c. In that case the initial bed level linearly increases from 9m –MSL at the basin mouth to MSL at the head. The applied numerical grid is shown in Figure 4.1a. A second investigated geometry has a basin length of 80 km. The final model setting allows for bank erosion of the 20 km and 80 km long basins. All shallow basins result in a tide residual export of sediment towards the sea. The initial basin mouth cross-sectional area below MSL is set at 1800 m^2. Friction is described by Manning's formulation with a coefficient of 0.026 sm$^{-1/3}$. Further, the grain size is defined uniform over the domain with a value of 240 μm.

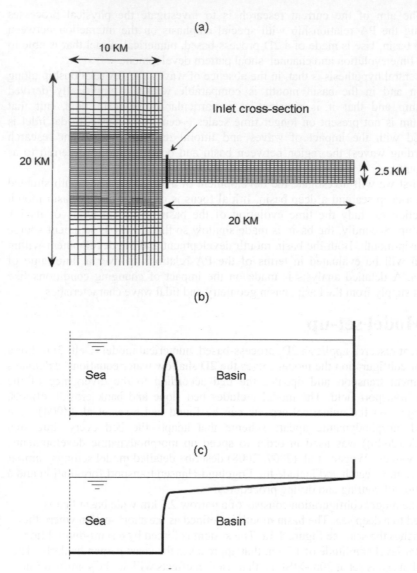

Figure 4.1 *(a) Computational grid of the process-based model; (b) cross-section of shallow basin mouth with deep basin; (c) cross-section of shallow basin mouth with shallow basin.*

4.3 Results

4.3.1 *Evolution of basin mouth connected to a deep and short basin*

The model shows morphodynamic evolution of the basin mouth over 1000 days. Figure 4.2a shows the basin mouth cross-sectional area development over time. The initial basin mouth profile appears to be too shallow and rapid erosion takes place, exporting sediments towards the sea during ebb and to the basin during flood. Over the longer term the morphodynamic development shows no equilibrium but ongoing, although continuously decreasing, increase of the profile. Following suggestions by Friedrichs (1995) and Gao and Collins (1994) spatial gradients in tide residual transports play a dominant role in the basin mouth evolution. The larger these gradients are, the faster the morphodynamic evolution of the cross-section will take place. Figure 4.3 shows that, with time, tide residual transports become spatially more uniformly distributed along the basin mouth, which leads to the decreasingly lower development rate (Figure 4.2a). In order to mimic wave induced sediment supply (littoral drift) towards the basin mouth sediment was added every time step into the basin mouth area. Figure 4.2b shows indeed development towards equilibrium, which is reached in about two years.

The initial basin mouth is too small for a given tidal prism compared to the empirical relationship suggested by Jarrett (1976). Morphodynamic development shows a continuous trend that somewhat slows down in the area close to the Jarrett relationship rather than a development towards a fixed equilibrium point (Figure 4.4(a)). The two lines represent the curves that Jarrett (1976) found based on empirical data of jettied and unjettied inlets. In case of extra sediment supply equilibrium is reached just between the two Jarrett curves (Figure 4.4(b)). The fact that the equilibrium is just between the two curves is merely coincidental. The equilibrium cross-sectional area will depend on the amount of sediment supplied compared to the tidal capacity to carry the sediment outside the basin mouth. The importance within the framework of the current study is that the results show a fixed equilibrium point only in case of extra sediment supply.

The next section deals with the consideration that, apart from wave induced sediment supply to the inlet, inlet evolution may also be a function of sediment supply from the basin side of the inlet. This sediment supply is determined by sediment availability in the basin and the morphology of the basin.

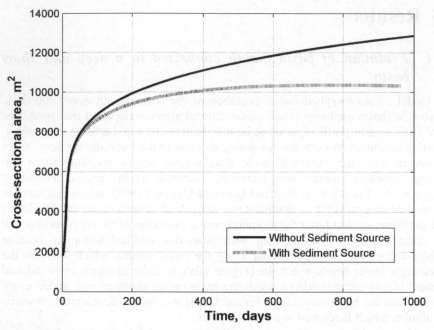

Figure 4.2 *Development of cross-sectional area over time with and without sediment source*

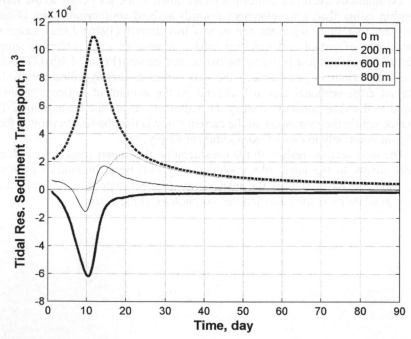

Figure 4.3 *Tide residual sediment transport in the first 90 days of simulation for cross-sections at a distance of 0, 200, 600 and 800 m landward from the basin mouth cross-section.*

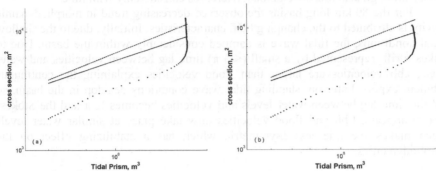

Figure 4.4 *Relation between P and A. Process-based model results compared to empirical Jarrett (1976) relationship - jettied (dash dot line) and unjettied (dashed line); Model results show 1000 day development. (a) Without extra sediment supply. (b) With extra sediment supply. In the latter case the cross-sectional area development stops somewhere between the Jarrett relations.*

4.3.2   *Impact of basin geometry on basin mouth evolution*

Description of model results

The following section investigates and analyses the impact of shallow basin geometry, which is capable of providing sediments to the basin mouth. Furthermore, the impact of long-term, changing basin bathymetry and tidal wave characteristics is analyzed within the context of the PA relationship. The analysis is based on model results in earlier research by Van der Wegen et al. (2007) and Van der Wegen et al. (2008), who present a description of the geometry and a detailed analysis of the hydrodynamic processes. Their results are summarized here for reasons of clarity.

The models start from an initially landward linearly sloping bed level similar to the one shown in Figure 4.1c. The initial bed is inherently not stable so that channel-shoal patterns develop along the basin first decades. These patterns become stable and gradually evolve on a much longer time scale. The initial bed level is too shallow for equilibrium causing the basins to continuously export sediments, albeit at a decreasing rate. Runs were carried out for 20 km and 80 km long basins with fixed banks (2.5 km wide) and for erodible banks starting from a width of 500 m. The runs reflect an evolution of about 3200 years. Figure 4.5 shows an impression of the results for the 80 km long basin. Van der Wegen and Roelvink (2008) compared the results of a similar model with the Western Scheldt estuary in the Netherlands and concluded that, despite the model simplifications, realistic results were obtained in terms of percentage of intertidal area, hypsometry and characteristic morphological wavelength.

The configurations with erodible banks show significant transport of sediment from the banks. This initially leads to a much shallower basin. On the longer term (~centuries), the models show continuous deepening as well. After millennia, developments are only minor in the fixed banks configurations. The erodible banks configurations show continuous widening and deepening of the basins, although also this development decreases considerably with time.

For the 80 km long basins the observed decreasing trend in morphodynamic activity is attributed to the changing tidal characteristics. Initially, due to the shallow initial conditions, the tidal wave is damped considerably within the basin. Due to Stokes' drift, represented by a small ($<\frac{1}{2}$ π) time lag between velocities and water levels, ebb velocities are larger than flood velocities explaining the continuous sediment export. Later on, standing tidal wave conditions develop in the basin, so that the time lag between water levels and velocities becomes $\frac{1}{2}$ π and the Stokes' drift disappears. Ebb and flood velocities now take place at similar water levels which makes the tide less asymmetric which has a stabilizing effect on the morphodynamics.

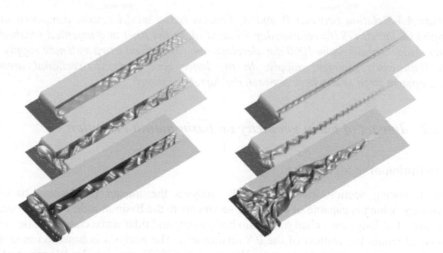

Figure 4.5 *Impression of bathymetry of 80 km long basin after 25, 400 and 3200 years for fixed banks (left) and erodible banks (right). Pictures reproduced from Van der Wegen et al (2008)*

Figure 4.6 shows PA relationships for the different model configurations for different moments in time along the basin. The 'O' symbol denotes values at the mouth. The tidal prisms were calculated by integrating the tidal discharge 'Q' over the cross-section 'A' from low water slack (LWS) to high water slack (HWS):

$$P = \int_{LWS}^{HWS} \int_{0}^{A} Q(A,t)\, dA dt \qquad (4.2)$$

Figure 4.7 gives an impression of the remarkably different cross-sectional geometries and sizes over time that all fit a tidal prism of $7*10^7$ m^3. A number of interesting observations can be made.

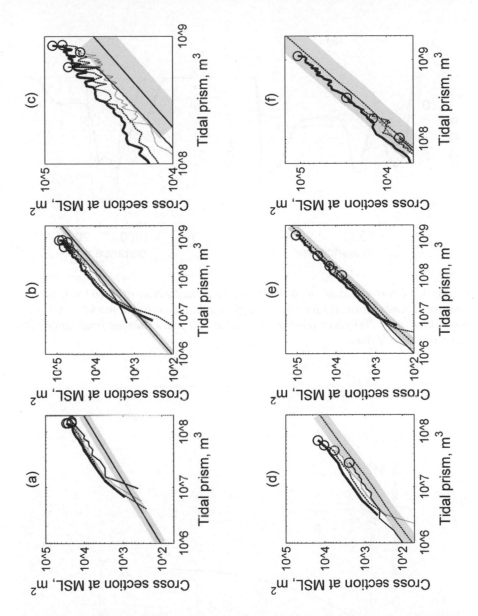

Figure 4.6 *Relation between (P) and (A) . (a) fixed short basin; (b) fixed long basin; (c) fixed long basin –detail; (d) erodible short basin; (e) erodible long basin; (f) erodible long basin - detail; Jarrett(1976) - jettied (Strait solid line); Jarrett(1976) - unjettied (Strait dashed line); gray area represents 95% confidence intervals (Jarrett 1976). Model results after 1 year (thin solid line); after 200 years (dotted line); after 800 years (thick dashed line); after 3200 years (thick solid line); open circles indicate values at the mouth.*

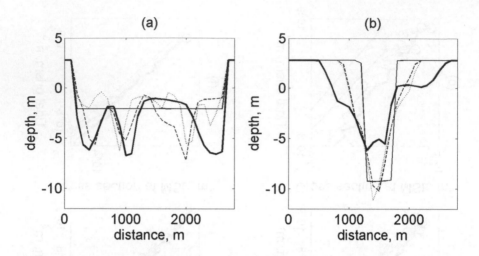

Figure 4.7 *Cross-sectional geometries that fit a tidal prism of 7*107 m3. (a) For long fixed banks basin; (b) for long erodible banks basin. Initial model results (thin solid line); after 200 years (dotted line); after 800 years (dashed line); after 3200 years (thick solid line);*

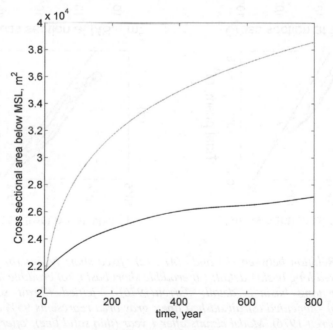

Figure 4.8 *Development of cross-sectional profile over time for basin mouth with deep basin (thin line) and a shallow short basin (thick line), both starting from a mean depth of 9 m below MSL.*

At the basin mouth

First observation is that model results at the mouth show similar values as the empirical PA relationship, although they somewhat overestimate the empirically found cross-sectional area for a given tidal prism. Time evolution shows that prisms and cross-sections increase at the mouth and that the cross-section area increases relatively more than the tidal prism. This is most pronounced in Figure 4.6d (short basin with erodible banks).

For the fixed banks configurations the PA relationship shifts merely upward to larger cross-sectional areas, so that, with time, the cross-sectional area increases for a given tidal prism. The upward shift takes place with a much lower rate than for the case that the basin mouth is attached to a deep basin. Figure 4.8 shows the development over time of a basin mouth attached to a deep basin and a shallow basin, both starting from a mean depth of 9 m below MSL. The two timescales can be clearly distinguished. Figure 4.9 shows a schematic representation of the processes responsible. In case of a deep basin velocities and associated transports are relatively high in the shallow basin mouth area. Consequently, the resulting spatial sediment transport gradients are high compared to a basin mouth connected to a shallow basin. Especially, the sediment supply from the shallow basin during ebb has a damping effect on the spatial transport gradients (and thus on the morphodynamic development).

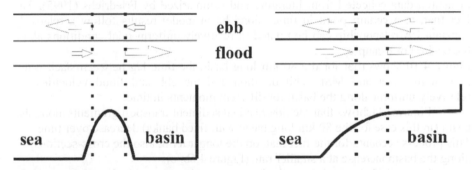

Figure 4.9 *Schematic representation of basin mouth with deep basin (left) and shallow basin (right). Arrows indicate transports during ebb and flood from and to the basin mouth area.*

Other processes may have a damping effect on basin mouth evolution as well. The presence of an ebb delta exports sediments towards the basin mouth during flood and reduces spatial gradients in tide residual transports (see for example Powell (2006)). River discharges will supply extra sediments as well, but they may also change flow characteristics in the basin mouth causing higher ebb velocities and lower flood velocities. These processes are considered outside the scope of the current research

Model results finally tend to overestimate the cross-sectional area belonging to a certain tidal prism, in particular for the fixed banks basins. By applying the Engelund-Hansen transport formulation, the current study does not include a critical shear stress for sediment motion. One could argue that this accounts for the continuous increase in cross-sectional area. However, runs with a Van Rijn transport formulation including a critical shear stress (not shown) lead to similar results. This suggests that, at least for this particular model configuration, gradients in transport are more important than the criterion of critical shear stress.

The fact that the model finally leads to larger cross-sections than suggested by the empirical PA relationship of Jarrett (1976) for jettied inlets may have several other causes. One reason is the highly schematized model setup. There is uniformity in friction, bed slope effects, grain size distribution and sediment compaction and non-erodible layers are excluded. The grain size in deep channels, for example, may be relatively large in reality due to prevailing high channel velocities. Grain sorting, armoring effects and sediment compaction at large depths would lead to shallower cross-sections. Another reason why model results overestimate Jarrett's (1976) relationship is that no littoral drift is taken into account. Although Jarrett (1976) considered jettied inlets, it is still possible that a significant amount of sediments, suspended by waves on beaches next to the jetties, may have been able to enter the inlet during flood.

Along the basin

The trend of the PA relationship along the basins (i.e. not only at the mouth) for different points in time shows a similar trend as the empirical PA relationship. This confirms observations along the Western Scheldt estuary by De Jong and Gerritsen (1984), Eysink (1990), Rinaldo et al (1999b) for the Venice lagoon and other measured data collected from literature and summarized by Friedrichs (1995). The fact that, at a certain point in time, along basin model results follow a relatively constant PA relationship can be related to relatively uniform tidal conditions along the basin. For example,

Figure 4.10 shows that for the 80 km long basin the time lag between slack water after high water and MSL, ebb duration and the ebb and flood velocities are relatively uniform along the basin for different moments in time.

Figure 4.11shows that the tide residual sediment transport gradients along the basin (in this case for the 80 km long basin with fixed banks) decrease over time. This process accounts for the fact that, on the longer term, also the cross-sections along the basin increase at a smaller rate (Figure 4.6b, c).

Figure 4.12 shows that, on the longer term, the tide residual spatial transport gradients are negligible on a basin mouth length scale, which suggests local equilibrium. The profile along the basin can thus be thought of as interconnected series of local (near) equilibria. This suggests that the constant PA relationship along the basin is a result of a nearly constant tide residual spatial transport gradient along the basin (Figure 4.11).

For the configurations with erodible banks both the prism and the cross-sectional area increase and there is a less pronounced upward shift. The reason for this is the dominant widening process of the basins that allows for larger tidal prisms. The widening process supplies sediment into the basin and prevents considerable deepening as in case of the fixed bank basins. For the long basin the widening process of the basin is ongoing with a timescale that is larger than the 3200 years run time. In case of the short basin the widening process decreases considerably after 3200 years, so that tidal prisms do not increase dramatically anymore.

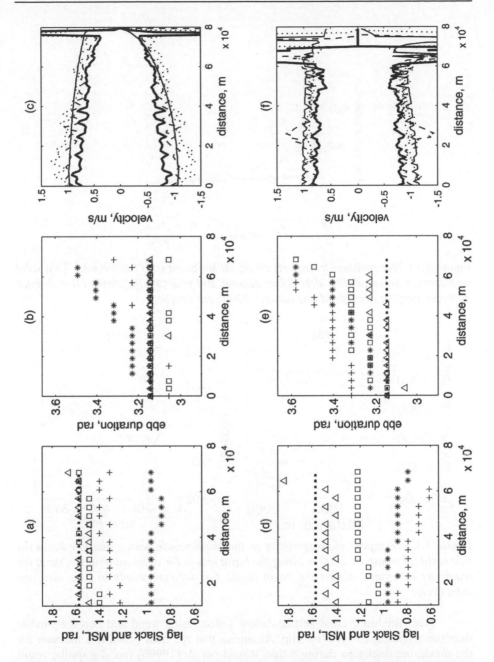

Figure 4.10 *The time lag between slack water after high water and MSL (a, d); ebb duration (b, e); and the ebb (negative) and flood velocities (c, f) for the long basin with fixed banks (a, b, c) and erodible banks (d, e, f). The irregular behavior near the head is due to the large percentage of intertidal area. '*' and thin solid line denote initial profile; '+' and dotted line denote 200 year profile; '□' and dashed line denote 800 year profile; '△' and thick solid line denote 3200 year profile;*

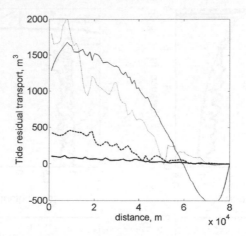

Figure 4.11 *Tide residual transports along 80 km basin with fixed banks. Thin solid line denotes initial profile; dotted line denotes 200 year profile; dashed line denotes 800 year profile; thick solid line denotes 3200 year profile;*

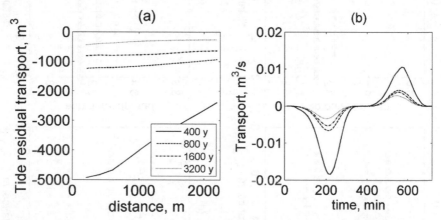

Figure 4.12 *Transport characteristics in the basin mouth with a shallow basin (a) tide residual transport 2000 m along the basin mouth for different points in time; (b) transport integrated across the basin mouth for different points in time over one tidal cycle;*

Near the head, cross-sections show a decreasing trend and become smaller than the empirical PA relationship. Assuming that the tidal prism is a measure for the maximum discharge during a tide, Rinaldo et al. (1999b) found a similar result for measurements in the Venice lagoon. A possible explanation for this can be associated with large tidal amplitudes, relatively low velocities and the presence of a considerable percentage of intertidal area per cross-section near the head (Figure 4.13). The cross-sections are defined by the area below MSL. It is assumed that, near the head and during rising water levels, a relatively large part of the hydrodynamic transport takes place across the intertidal area, which would additionally allow for smaller channel cross-sections. During ebb a similar process occurs, until the intertidal area falls dry again.

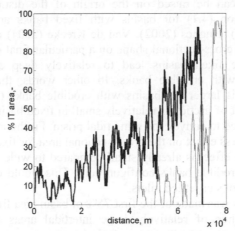

Figure 4.13 *Percentage of intertidal area per cross-section for long basin after 3200 years; fixed banks (thin line) and erodible banks (thick line).*

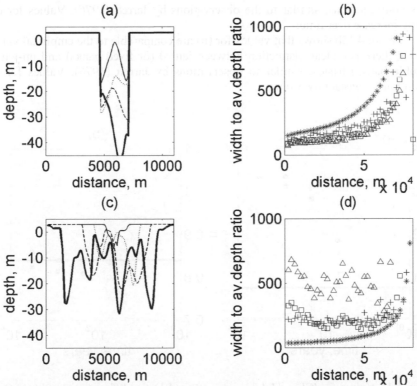

Figure 4.14 *Different cross-sectional profiles along the 80 km long basin after 3200 year. (a) For fixed banks; (c) for erodible banks. Thin solid line denotes profile at 60 km; dotted line denotes profile at 40 km; dashed line denotes profile at 20 km; thick solid line denotes profile at mouth; Width to width-averaged depth ratios along the basin (b) for fixed banks; (d) for erodible banks. '*' denotes initial data; '+' denotes 200 year data; '□' denotes 800 year data; '△' denotes 3200 year data;*

The question can be raised on the origin of the distinct values for the coefficients in equation (4.1) for basins with fixed banks and erodible banks. Gerritsen et al. (1990), Hughes (2002), Van de Kreeke (2004) and Walton (2004) point to the impact of cross-sectional shape on a particular final equilibrium. Model results show that the fixed basins lead to relatively deep and narrow basins compared to basins with erodible banks. In other words, the width to width-averaged-depth ratio is larger for basins with erodible banks (Figure 4.14). This implies that the impact of friction is relatively small in fixed basins so that a smaller cross-section is required to convey the same tidal prism. Furthermore, an increase in tidal prism will have less effect on the cross-sectional area for fixed banks due to the fact that the frictional effect is already small compared to wide and shallow cross-sections (like in the erodible banks configuration). This would explain the smaller (n) values for fixed banks or jettied inlets.

Closer analysis was made of the first 75% of basin area from the mouth, thus neglecting the influence of relatively large intertidal areas near the head. A regression analysis was made by curve fitting an exponential relationship like equation (4.1) to the data. Figure 4.15a shows that model results overestimate values for (C), although there is a clear distinction between the fixed banks and erodible banks configurations, similar to the observations by Jarrett (1976). Values for (C) slightly increase over time.

Figure 4.15b shows that values for (n) are comparable to the empirical values and that there is a clear distinction between jettied (or fixed banks) and un-jettied (erodible banks) basins, similar to observations by Jarrett (1976). Values for (n) remain fairly constant over time.

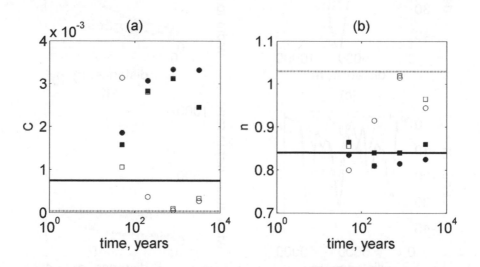

Figure 4.15 *Values for 'C' and 'n' from equation (4.1) for different points in time. '■' Denotes short basin with fixed banks; '□' denotes short basin with erodible banks; '●' denotes long basin with fixed banks; '○' denotes long basin with erodible banks; solid line represents Jarrett (1976) for jettied basin mouths; dashed line represents Jarrett (1976) for un-jettied basin mouths.*

4.3.3   Time scales and sea level rise

The PA relationship found in the current research does not describe equilibrium, but rather relatively constant conditions due to very small spatial gradients in tide residual transport. Because of the importance of these gradients, the current research suggests that inlet evolution depends on adjacent morphodynamic features that act as a sediment source and reduce these transport gradients, such as ebb and flood delta's or a shallow basin.

The empirically derived PA relationship by Jarrett (1976) does not explicitly show the impact of inlet evolution timescales. Model results suggest that ongoing, relatively small adaptations in basin mouth cross-section are of similar order of magnitude as the measured 95 % confidence interval given by Jarrett (1976). Jarrett (1976) partly explains the scatter by discriminating between jettied and unjettied inlets and between location of different inlets with different wave and/or tidal conditions. Also Jarrett's computational methods to calculate the prisms may account for some of the scatter. The current research suggests that time evolution processes and the age of basin mouths may be partly responsible for the scatter in empirical data as well.

The model results are based on constant hydraulic forcing conditions and do not take into account sea level rise and associated changes in tidal wave characteristic at the mouth and in the basin. Fleming et al. (1998) conclude that, after a period of rapid sea level rise, sea level has been relatively constant over the past 6000 years, although they do not exclude considerable fluctuations during this period. Considering the long time scale of millennia for adaptations of the profile, sea level rise may indeed be relevant on the long term.

Also, future rise in sea level (estimated at 30-80 cm for the coming century for the Dutch coast) may have considerable impact on the inlet and basin morphodynamics. The impact of sea level rise on the PA relationship is not straightforward. Higher water levels will increase the cross-section below MSL, but may also lead to a larger tidal prism. The impact will be different for different estuaries, depending amongst others on the availability and characteristics of sediments, changes of tidal wave characteristics and the geological history the basin.

4.3.4   Interaction between tidal channels and inlets

The physical explanation of PA relationships in tidal channels will probably be different from the explanation in inlets which experience impact of littoral drift. Literature review and model results showed that equilibrium in inlets can be explained by the external input of sediments such as littoral drift, whereas the current research suggests that, even on long time scales, no equilibrium develops in tidal channels. In most cases the littoral drift will also largely exceed the contribution of tide residual transports from the basin, making the latter process insignificant in the inlet. Equilibrium develops in the inlet, whereas the adjacent tidal channels (very) slowly deepen. Inlet cross-section will thus probably be shallower than tidal channel cross-sections. Features such as flood deltas will experience transient conditions.

4.4 Conclusions

The current research shows that a process-based, numerical modeling approach is able to describe long time evolution of a tidal basin under highly schematized conditions. The PA relationship in the basin mouth and in the basin compares well with the empirically derived PA relationship by Jarrett (1976). However, on long (~century) time scales model results show increasing cross-sections for similar tidal prisms although the deviations are comparable with the 95% confidence interval found by Jarrett (1976). Similar to the distinct PA relationships for jettied and unjettied inlets suggested by Jarrett (1976) the current research observes also distinct values for basins with fixed banks and erodible banks.

In the absence of waves and other sources of sediment supply the cross-section of a shallow basin mouth attached to a deep basin rapidly increases. In terms of the PA relationship this means that first both the cross-section and the tidal prism increase and that, in later stages, the basin mouth shows cross-sectional expansion under a constant tidal prism. The expansion takes place at increasingly smaller rate which can be related to decreasing spatial gradients in tide residual transports. Extra supply of sediments into the basin mouth leads to equilibrium of the basin mouth cross-sectional area. This extra supply may for example result from littoral drift as well as from a basin attached to the basin mouth.

The presence of a shallow basin has a damping effect on the basin mouth evolution. On a decadal time scale the process-based approach seems to reproduce a stable PA relationship, which shows good resemblance to the empirical PA relationship suggested by Jarrett (1976). However, on the longer term (~centuries) the PA relationship changes in the sense that cross-sections increase for similar tidal prisms.

The observation that the model results compare well with the PA relationship along the basin, is attributed to constant tidal wave characteristics. Small, tide residual transports gradients along the basin allow for the development of local, near-equilibrium. Similar to the empirical PA relationship by Jarrett (1976), model results can discriminate between confined (or jettied) and 'free' (or single jettied) inlets. Explanation is found by the fact that fixed banks lead to relatively narrow and deep cross-sections so that the impact of friction is relatively small. Jettied inlets thus require smaller cross-sectional areas than free inlets to convey the same tidal prism. Values of (n) remain fairly constant, whereas values of (C) show significant increase over time.

The current research suggests that the PA relationship describes relatively constant conditions (governed by dominant morphodynamic forcing mechanisms), while, at the same time, these conditions allow for slow modification of the PA relationship over time. The PA relationship found in the current research does not describe equilibrium, but rather relatively constant conditions due to very small spatial gradients in tide residual transport. Because of the suggested importance of these gradients, the current research suggests that basin mouth evolution should be studied in close relation to the surroundings of the basin mouth, especially when littoral drift is absent or small. Relevant morphodynamic features in this sense are sediment sources like ebb and flood delta's or shallow basins.

4.5 Acknowledgements

Preliminary discussions on the contents of this article with Henk Jan Verhagen (TU Delft) and Ko van de Kreeke (University of Miami) highly contributed to the quality of this chapter.

5 Reproduction of the Western Scheldt bathymetry by means of a process-based, morphodynamic model[4]

Abstract

Channel-shoal patterns in alluvial estuaries evolve from the interaction between tidal movement, the available sediments and the geometry of a tidal basin. Once such a system is not disturbed too much by extreme events, often a kind of morphological equilibrium seems to be present which describes only slow development on a decadal time scale. Earlier research shows that typical morphological length scales can be related to the tidal excursion. In case that the geometry is fixed by bank protection or non-erodible substrate (i.e. rock or peat), it directs the tidal flow and plays an important role in the allocation of channels and shoals.

The current chapter aims to investigate the importance of the geometry on the allocation and evolution of channel-shoal patterns in the Western Scheldt, the Netherlands. Use is made of a 3D morphodynamic, process-based numerical model (Delft3D). Starting from an initially flat bed, model results show morphodynamic development over 200 years towards reasonably stable channel-shoal patterns which represent major morphological characteristics of the Western Scheldt estuary. We applied different indicators and techniques to evaluate the model performance over time, namely visual comparison, hypsometries, Brier Skill Score (BSS), mean basin depth and longitudinal profiles. The most advanced indicator would be the BSS, although additional analysis is required to clarify observed patterns and processes.

Sensitivity analysis shows that trends in development are similar for different model parameter settings, like introducing salt-fresh water density differences or variations in sediment grain size. Varying river discharge and including non-erodible layers and dredging and dumping activities led to occasional improvements. 3D approaches led to better results than 2D. Both the Van Rijn and Engelund-Hansen sediment transport formulations lead to good results. It is hard to determine which of the cases has 'best' settings, because skill of the runs differs over time and even combining best model settings in a new run does not automatically lead to a better performance.

Main conclusion of the study is that the interaction between tidal forcing and basin geometry plays a significant role in the evolution and allocation of major morphological features in an alluvial tidal basin such as the Western Scheldt.

---

[4] A slightly adapted version of this chapter will be submitted to Geomorphology by authors Mick van der Wegen and Dano Roelvink

5.1 Introduction

5.1.1 Scope

The morphodynamic behavior of estuaries is determined by the complex interplay between hydrodynamic movement, sediment transports and bed level adaptations. Processes and phenomena of different spatial scales and time scales interact with each other in a non-linear manner. Besides regular tidal movement, examples of hydrodynamic processes are waves, density currents, discharge regimes or extreme events like storms causing water level setup and large waves. Sediment transport is further determined by the presence and spatial distribution of mud and/or sand and their specific characteristics at the surface, but also within the bed. Furthermore, the bed characteristics may be influenced by benthic flora and fauna and can be a function of biological processes like the development of bio films or the presence of bio engineers. All these processes may be a function of time so that they may change considerably over months, over the year or over decades. Considering this list of processes and seemingly required data, reasonable prediction of morphological development in an estuarine environment may look like an impossible task.

On the other hand, typical and similar morphological features can be recognized in different alluvial estuaries. For example, channel-shoal patterns have characteristic length scales that relate to the tidal excursion. This shows that main morphodynamic features can be related directly to dominant hydrodynamic (tidal) forcing conditions and suggests that other, less pronounced processes mainly follow the main morphology.

5.1.2 Aim of the study

Process-based numerical models have evolved last decades towards robust tools to predict morphodynamic developments based on detailed descriptions of relevant processes and advanced morphodynamic update schemes. As such, these models are excellent tools to investigate characteristic morphodynamic features developing on an alluvial bed. Following this approach, Hibma et al (2003), Van der Wegen et al (2007, 2008, 2009) and Van der Wegen & Roelvink (2008) present results of stable channel-shoal patterns emerging in elongated embayments starting from an initially flat bed under highly schematized tidal forcing conditions. Also the geometry of these basins was strongly simplified.

As a follow up on these investigations one may ask the question what the impact would be of the geometry on the development of characteristic morphological length scales and on the allocation of the main morphological features. A possible way of investigating this is to prescribe the geometry of a real estuary and to start morphodynamic runs from a flat bed. Furthermore, by comparing the model outcomes to a real bathymetry an assessment can be made of the main forcing mechanisms, processes or sediment characteristics that determine the morphodynamics within the given estuarine geometry.

The current study aims to investigate the skill of morphodynamic, process-based numerical models in reproducing realistic bathymetries and in the absence of detailed data.

5.1.3   Approach

As explained in the previous section a process-based numerical model (Delft3D) will be applied that is able to handle different forcing conditions and different (sediment) parameter settings for a given estuarine geometry.

The assessment of model skills requires a clear methodology, which makes objective comparison to model parameter variations, other case studies or other modeling approaches possible. Visual comparison of modeling results to a real bathymetry can be excellent as a first guess on the model performance, but it lacks an objective criterion. Modeling results are therefore analyzed in terms of the Brier Skill Score (BSS). Also comparison will be made between measured and modeled hypsometries and volumetric changes.

5.1.4   Case study

Focus of the study is on the Western Scheldt, the Netherlands. The main reason to choose this estuary is that tidal hydrodynamics dominates river discharge and that river sediment supply is almost absent. The morphology is thus mainly determined by tidal water movement. A second reason is that much data on this estuary is available so that systematic assessment can take place of the dominant forcing mechanisms. Another advantage is that the bed mainly consists of sand, although mud is found on intertidal areas and salt marshes.

The Western Scheldt estuary is located near the Dutch Belgian border at about 51^0 latitude. It is a relatively young coastal plain estuary that developed since the early Middle Ages when a tidal channel penetrated landward towards the Scheldt River, located north of Antwerp, initially developing over a bed of marshland (Van der Spek (1997), Beets and Van der Spek (2000)). Alluvial sand and remnants of erosion resistant peat layers determine the present bathymetry. Land reclamation taking place over several centuries resulted in dikes constructed on the banks of the Western Scheldt, including scour protection at the bed sometimes extending into the deepest channels adjacent to the dikes (Toffolon and Crosato (2007)).

5.2 Model set up

The Delft3D hydrodynamic model and the morphodynamic model have been described in detail in Chapter 1.

A curvilinear grid as depicted in Figure 5.1 was created for this study. The grid extends about 20 km seaward from the coastline and landward almost up to Gent. The grid has typical grid cell sizes of approximately 100 by 200 m in the central area, extending to several km's near the sea boundaries. In the 3D runs, 10 sigma layers describe the vertical grid with increasing resolution towards the bed. The time step is 1 minute.

The seaward boundary conditions were generated by nesting this detailed Western Scheldt model into a larger and coarser model covering large part of the North Sea, which was driven by astronomical components. As in Roelvink and Walstra (2004), the model applies a combination of a water level boundary conditions (at the north-western seaward boundary) and Neumann boundary conditions (at the south-western and north-eastern seaward boundaries). In the latter boundary condition, the alongshore water level gradient due to phase and amplitude gradients is prescribed, which allows the water level and flow profile along the lateral boundaries to develop freely without disturbances. The landward boundary prescribes a constant discharge, without sediment supply.

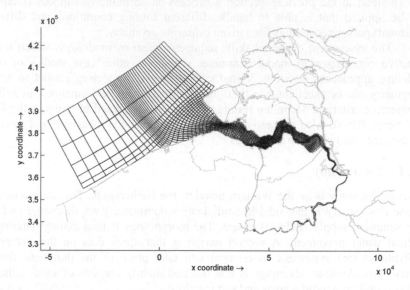

Figure 5.1 *Model grid (black) and land boundary (orange)*

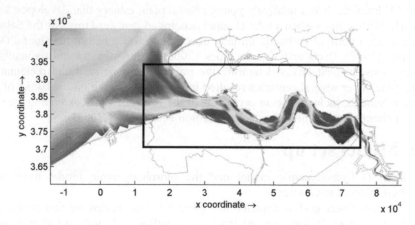

Figure 5.2 *1998 bathymetry and focus area of current study (outlined)*

The initial bed level was derived from the measured 1998 bathymetry. The level was averaged over the domain outlined in Figure 5.2, so that the initial bathymetry describes a horizontal bed bordered by an ebb delta in the west and a river type of geometry in the east.

The approach of the current research is to start from a simple model, excluding as many processes as possible. Systematic further runs are applied so that dominant processes and parameters can be revealed. A total of about 30 runs were carried out. A typical (3D) run duration is about a day on a decent PC. It covers half a year of hydrodynamic simulations similar to 200 years of morphodynamic evolution applying a morphological factor of 400.

The standard case applies the Engelund-Hansen transport formulation in a 2D setting, a lateral bed slope factor (α_{bn}) of 10, a sediment layer of 50 m deep, a grain size (D) of 200 μm, a river discharge of 15 m^3/s, and M_2, M_4, M_6 and C_1 tidal components as sea boundary conditions (with amplitudes of 1.72m, 0.09m, 0.05m and 0.17m respectively). The C_1 tidal component is an artificial diurnal component derived from the combination of the O_1 and K_1 tidal components. The derivation is based on the assumption of sediment transport similarity and has the advantage that it does not include a spring-neap tidal cycle due to the small differences in period between O_1 and K_1. Hoitink et al. (2003) and Lesser (2009) give further details on the derivation of C_1. The Van Rijn sediment transport formulation (Van Rijn, 1993) makes distinction between bed load and suspended load. The bed slope factor only affects the bed load. Realistic results were only obtained by setting the value of (α_{bn}) an order of magnitude larger that in case of the Engelund-Hansen transport formulation. The van Rijn bed load is a function of local velocities, whereas the suspended load is calculated based on an equilibrium sediment concentration and the advection-diffusion equation. It thus allows for settling and suspension time lags. A list of abbreviations of different model runs is given in Table 3. The standard case is referred to as EH-2D.

Parameter	*Abbreviation*
Engelund-Hansen transport	EH
Van Rijn transport	vR
Two dimensional	2D
Three dimensional	3D
Including non-erodible layers	nel
Including dredging and dumping activities	d&d
Bed slope factor X	abnX
River discharge 500 m3/s	Q500
Only M2 tidal component	M2

Table 3 *Abbreviations applied for names of different runs*

The Western Scheldt includes areas consisting of sediment layers that are hard to erode like peat layers and Pleistocene. These layers can be included in the model by adjusting sediment availability per grid cell. Zero sediment availability defines a non-erodible layer so that no erosion is possible on that location, unless new sediments deposit.

The Western Scheldt is subject to extensive dredging activities to maintain sailing depth towards the Port of Antwerp in the hinterland. These dredging activities may also be included in the model by defining areas in which a certain depth is maintained. Once sediment deposits in this area at a certain time step it is immediately removed and allocated in another predefined area. This area may be situated outside the modeling domain but also within the domain. This latter option is applied since dredging policy is to keep sediment within the Western Scheldt system. For the dredging areas the depth is maintained at 12.5 m.

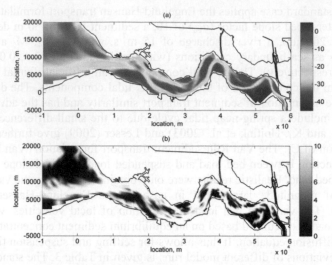

Figure 5.3 *(a) measured 1998 bathymetry; (b) indication of location of non-erodible layers in meters of available sediments*

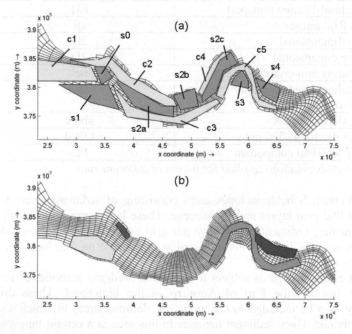

Figure 5.4 *(a) Schematization of main morphological features of 1998 bathymetry, c=channel (light blue), s=shoal (orange). (b) Dredging areas (orange and red) and dumping areas (light and dark blue). Following roughly dredging policies sediment from red area is deposited in dark blue area and sediment from orange area is deposited in light blue area.*

5.3 Model results

5.3.1 General observations

A total number of about 30 runs was carried out. The results are not all reported here, mainly because some of the differences were not considered significant compared to the standard case results. These runs included variations of the grain size between 80 μm and 300 μm, varying the morphological factor from 50 to 400, and including or excluding salt/fresh water density differences.

Figure 5.5 shows an example of the typical modeled morphodynamic evolution of the basin, namely case (EH-3D) which is the run performing the best in terms of the Brier Skill Score discussed in more detail in the following sections. The initial morphological developments (after 15 years) are relatively fast and show immediate deepening in the narrow mouth section and around land boundary outcrops. Major shoals develop east of the mouth section. The morphological features after 30 years are already comparable to the 1998 bathymetry in location and size, although some channels and shoals are less pronounced in the model. Morphodynamic activity decreases on longer time scales although development continues to take place.

5.3.2 Visual comparison of pattern development

Figure 5.6 shows bathymetries for different parameter settings after 200 years of morphodynamic evolution. Figure 5.6 (a) is the measured 1998 bathymetry. With reference to the main morphodynamic features (Figure 5.4 a) one can observe some remarkable resemblances, but also significant differences. All cases clearly show development of a flood channel penetrating the estuary from the sea. This channel develops along the southern banks seaward from the mouth, which can be explained by the tide that propagates from the south (Dissanayake et al. 2009). In most runs the penetrating channel is formed by merging c1 and c2 channels. South of the penetrating channel forms a more bended ebb channel (c3). More landward channel c5 develops. Channel c4 is not very pronounced, but always present. 2D runs show channels that are quite straight (Figure 5.6 b,c,d), whereas 3D runs show more bended channels, resulting from secondary flow excavating bends. A low value of the bed slope factor (α_{bn}) results in more channels in lateral direction that are relatively narrow and deep, which seems unrealistic (Figure 5.6 c). Van Rijn runs (Figure 5.6 d,j) seem to result in more diffused patterns due to the presence of settlement and suspension time lags, but this can also be attributed to the effect of the bed slope factor on the resulting sediment transport, which differs from the effect in case of the Engelund-Hansen transport formulation. 3D Results (Figure 5.6 e-j), generally lead to a better visual fit to the measured morphology. The run including the M2 tidal signal only (Figure 5.6 f) already leads to quite significant resemblance with the observed bathymetry. Including a non-erodible layer (Figure 5.6 g) shows maybe the best visual fit to the measured bathymetry. All shoals except s0 can now be clearly distinguished. Including the dredging and dumping activities (compare Figure 5.6 g and h) leads to a higher s1 shoal and lower bed levels landward of s4. The Q500 run (Figure 5.6 i) shows a more pronounced c3 ebb channel than the run without river discharge (Figure 5.6 h). This is probably caused by the fact that a high river discharge enhances ebb velocities and ebb channel (c3) development.

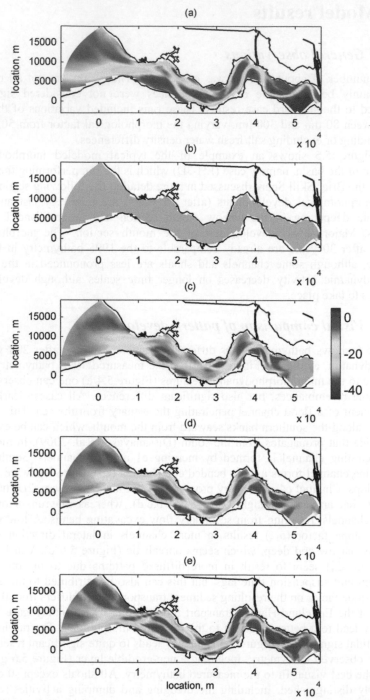

Figure 5.5 *Bed level development of EH-3D case. (a) measured 1998 bathymetry; (b) initial bathymetry; (c) bathymetry after 15 years; (d) bathymetry after 30 years with best BSS; (e) bathymetry after 200 years.*

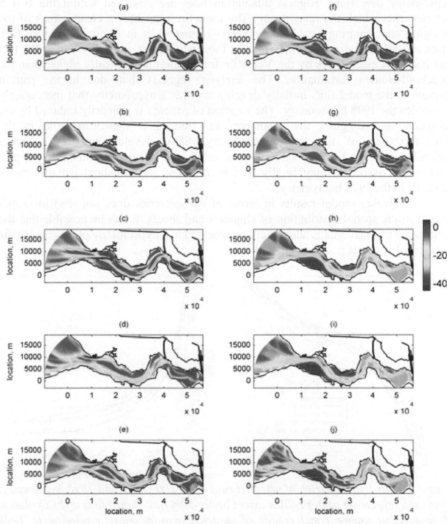

Figure 5.6. *Bed level development after 200 years after starting from a flat bed. (a) measured 1998 bathymetry; (b) EH-2D; (c) EH-2D-abn5; (d) vR-2D-abn100; (e) EH-3D; (f) EH-3D-M2; (g) EH-3D-nel; (h) EH-3D-nel-d&d; (i) EH-3D-nel-d&d-Q500; (j) vR-3D-abn100-nel-d&d.*

5.3.3 Comparison of hypsometries

A hypsometric curve provides information on the percentage of bathymetric area that is situated below a certain bed level. It provides, for example, information on the percentage of intertidal area and the channel area within a domain. Hypsometric plots for the same cases as in the previous figure are given in Figure 5.7a. Figure 5.7b shows time evolution of the root-mean-square (rms) errors of the different hypsometries compared to the 1998 hypsometry.

The figures show that all cases lead to an increasingly better fit to the 1998 hypsometry over time. Highest adaptation rates are observed within the first 5 decades after which the rates decay. The model results show a clear impact of the dredging and dumping activities between -10 and -5 m in Figure 5.7a. Van Rijn cases show fastest decay in rms value in the first decades, probably due to the fact that transport magnitudes by the Van Rijn formulation are typically higher than the Engelund-Hansen formulation. The analysis suggests that, due to the pattern formation, the model runs initially develop towards a hypsometry that increasingly resembles the 1998 hypsometry. The location of patterns is primarily induced by the geometry and suggests that geometry has indeed a significant role in the characterization of the basin bathymetry. Slow developments in sediment redistribution over the domain (such as small, tide residual, landward transport induced by overtides) may finally lead to different channel-shoal patterns and a better fit to the 1998 bathymetry.

Analyzing model results in terms of hypsometries does not lead to proper insight in the spatial distribution of channels and shoals. It can be possible that the percentage of shoal area is adequately represented in a hypsometric plot, whereas the location of shoals is poorly predicted.

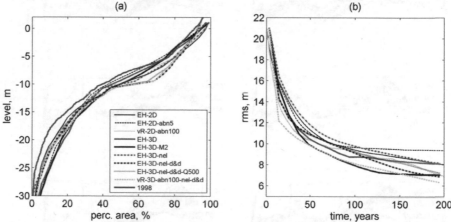

Figure 5.7 *(a) Hypsometries of selected runs after 200 years and 1998 hypsometry. Mean sea level is set at 0. Profiles extent up to -50m (not shown fully); (b) Evolution of root mean square (rms) errors of modeled hypsometries in relation to 1998 hypsometry.*

5.3.4   Brier Skill Score (BSS)

Visual comparison of model results to measured data and further process analysis give some insight into the model performance. Nevertheless, this methodology lacks objectivity. Sutherland et al (2004) suggest several other methodologies to assess the performance of morphodynamic models. In particular, they suggest the application of the Brier Skill Score (BSS). Chapter 1.4.5 describes the BSS in detail. The BSS addresses model performance compared to measured developments with respect to an initial bed (or: 'baseline' prediction). In the strict sense, the current study does not have a 'baseline' prediction. The study does not assess model performance over time (reproducing for example the morphodynamic evolution from 1998 to 2005), but aims to reproduce a bathymetry at a particular point in time. However, by taking the initial flat bed as 'baseline' prediction, model assessment in terms of the BSS becomes possible and provides an objective tool to compare model performance under different model parameter settings. The volumetric changes referred to in equation (1.15) are the volumetric differences with respect to a flat bed. The level of this flat bed is defined by the measured 1998 bathymetry averaged over the domain outlined in Figure 5.2b.

Figure 5.8 a,b gives an overview of the BSS of selected model runs over 200 years. First decades, all cases show positive BSS's as well as a BSS peak after which the BSS declines to negative values in most cases indicating that the modeled bathymetry is assessed worse than the initial flat bed. It is remarkable that, based on visual observations only, one would argue that the bathymetry after 100 year does not differ much from the bathymetry after 200 years (compare Figure 5.5 d and e), whereas the BSS after 200 years for the EH-3D case is much worse. Only cases including non-erodible layers and case EH-2D show a BSS increase again on the longer term. For the EH-3D-nel case BSS becomes even higher than the first decades BSS peak.

The vR runs show peaks earlier than the EH runs. Apart from the different nature of the vR formulation with a subdivision in bed load and suspended load, this is mainly attributed to the generally larger volumes transported by vR calculations or the value for (α_{bn}). The vR peaks are lower than EH peaks. On the long term 3D cases perform generally better than 2D cases. The case with high river discharge finally performs better than cases with low river discharge. The case with M2 only leads to worse results than including the full tidal forcing, although performance is better on the longer term.

The development of, for example, the EH-3D case (in red) can be understood by looking at the BSS decomposition parameters of phasing (α), amplitude and phasing (β) and volume (γ), see Figure 5.8 c,d and e. The analysis shows that the strong initial performance of case EH-3D can be attributed to improved phasing (α). The longer term decay in BSS is a result of worse fit in volume (γ) and amplitude and phasing (α,β). The worsening (β) value in the first decades can be attributed to amplitude performance, since phasing (α) still improves in that period. On the longer term BSS decays because of worse results in all three parameters. The long term increasing BSS for case EH-3D-nel can be attributed fully to phase improvement. Volume (γ) performance decreases and the combination of amplitude and phasing (β) remains fairly constant. Under conditions of phase improvement, a constant (β) value implies that amplitude performance decreases.

The BSS values are quite high. Following the classification by Sutherland et al. (2004) best results are considered reasonable (0.1-0.2) to good (0.2-0.4). It is questioned, however, whether model results can be categorized according to this classification, since the baseline prediction is a flat bed and not a measured bathymetry. As a result, the measured volumetric differences are already quite high which automatically leads to a high BSS. On the other hand, the BSS analysis shows positive scores, which means that the model significantly improves the morphology compared to a flat bed. Also, the BSS provides a good tool for inter-comparison of the different cases.

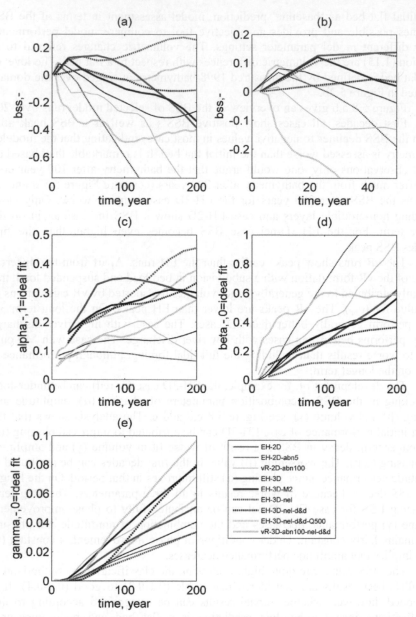

Figure 5.8 *Model results over time for selected runs (a) BSS; (b) detail of (a); (c) α, 1=ideal fit; (d) β, 0=ideal fit; (e) γ, 0=ideal fit;*

Figure 5.9 shows that all cases show a deepening model domain. Closer analysis reveals that this is mainly due to seaward export. The EH-3D-M2 case shows highest export of all EH cases, which is also supported by Figure 5.8 e. This is probably due the absence of M4 and M6 tidal components that would favor sediment import (or: less export). In case of EH a 3D run and dredging and dumping activities lead to a deeper basin.

Figure 5.10 compares the width-averaged, longitudinal profiles of 4 selected runs to the measured longitudinal profile. All runs show deeper profiles than the 1998 bathymetry, but also resemble the measured profile with depth peaks around 27, 38 and 60 km and shoal around 32, 40, 58 and 62 km onwards from the mouth. These peaks can be linked to the geometry. The depth peak around 27 km, for example, is related from the narrow cross-section at the mouth. Dredging and dumping activities lead to sediment redistribution with higher bed levels between (roughly) 30 and 45 km and deeper profiles from 62 km onwards.

Finally, Figure 5.11 shows the BSS of the width averaged depth (from Figure 5.10) over time. It can be clearly seen that the cases with dredging activities show behavior distinct and worse from the other cases. This holds for all three parameters (α, β and γ) in the BSS (not shown) and suggests that the dredging and dumping applied in the model needs improvement. Especially, the amount dredged from 62 km onwards needs a better schematization.

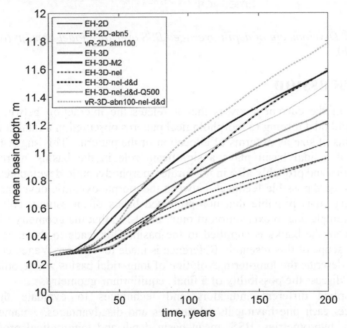

Figure 5.9 *Development of depth, averaged over the model domain (outlined in Figure 5.3b).*

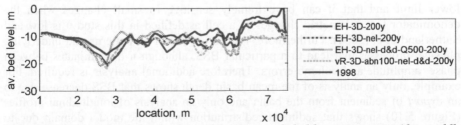

Figure 5.10. *Averaged depth along the Western Scheldt estuary axis (the middle between the two banks) for different runs. Origin is at the mouth. Depth was averaged over width perpendicular to axis, which is roughly perpendicular to the banks.*

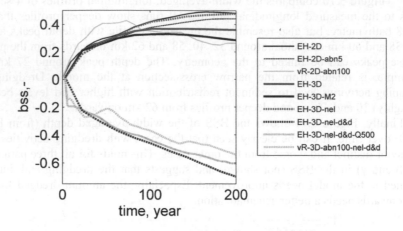

Figure 5.11 *Development of depth-averaged BSS over the model domain (outlined in Figure 5.3b).*

5.4 Discussion

The value of the current study is that it relates the interaction between a fixed geometry and tidal forcing to morphological patterns observed in a tidal embayment, both in terms of size as in terms of allocation of the patterns. This suggests that the geometry of a tidal basin plays a significant role in the basins' morphological characteristics and probably also in its further morphodynamic development.

We disregard possible feedback between the morphodynamic developments and the geometry. It is possible that non-protected banks of an estuary collapse over time, for example due to excavation of outer bends, so that the geometry widens and sediment from the banks is supplied to the basin. This leads to dynamics that are outside the scope of this research. Reference is made to Van der Wegen et al (2008, 2009) who describe the long-term evolution of long tidal basins with erodible banks and further discuss the possibility of a final 'equilibrium' geometry.

We applied different indicators and techniques to evaluate the model performance, each one having its advantages and disadvantages, namely visual comparison, hypsometries, BSS, mean basin depth and longitudinal profiles. The most advanced indicator would be the BSS that includes most objective information on model performance. The disadvantage of the BSS is that it is unbounded at the lower limit and that it can be extremely sensitive to small changes when the denominator (or: volumetric changes in a cell as defined in this study) is low and Sutherland et al (2004). Furthermore, it does not reveal all patterns and underlying physical processes that lead to a particular BSS, although it discriminates between phase, amplitude and volume errors. Therefore additional analysis is required. For example, only an analysis of the mean basin depth shows that BSS decrease due to an *export* of sediment from the basin and only an analysis of longitudinal profiles (Figure 5.10) shows that sediment redistribution within the model domain due to dredging and dumping activities may be responsible for the improving BSS scores on the longer term (Figure 5.8a) despite the increase of mean basin depth (Figure 5.9).

The model results show similar trends under variations of model parameter settings. It is hard to determine which of the cases has 'best' settings, because performance of the runs differs over time. Even combining best model settings in a new run docs not automatically lead to better performance. It is important to note, however, that trends in development are similar in many cases like the increase of the mean basin depth or the development of a BSS peak within decades. Including mud transports or more processes like wind waves, sand-mud interaction or biological process, may enhance the model performance.

By starting from a flat bed we violate the original and historic evolution of the Western Scheldt estuary. This makes the model results even more remarkable. It suggest that the evolving patterns have a timescale that is captured within the modeled period. The BBS peak reached within decades gives a good indication of the time scale of pattern development. Model trends, such as the deepening of the width averaged profile, have a much longer time scale and although it can be questioned whether or not they reflect historic or future developments in the Western Scheldt basin. The observation that all cases describe a net sediment export from the basin may be a consequence of the initially flat bed and the resulting bathymetry, but it may also be a proper description of the long-term trend in the Western Scheldt system. This long-term trend is subject of current discussion. Measurements suggest that the Western Scheldt changed from an importing to an exporting system during the period 1980-1990 (Nederbragt and Liek (2004) *in Dutch*). The reasons for this are unclear and are subject of current research, although Nederbragt and Liek (2004) *in Dutch* suggest a the considerable influence of dredging and dumping operations on the decadal sediment balance. This suggests that the long-term morphodynamic behavior is a delicate process to model and that morphodynamic reproduction may significantly depend on subtle model settings, boundary conditions and processes like wind waves or including mud transports.

5.5 Conclusions

Channel-shoal patterns in alluvial estuaries evolve from the interaction between tidal movement, the available sediments and the geometry of a tidal basin. Once such a system is not disturbed too much by extreme events, often a kind of morphological equilibrium seems to be present which describes only slow development on a decadal time scale. Earlier research shows that typical morphological length scales can be related to tidal wave characteristics. In case that the geometry is fixed by bank protection or non-erodible substrate (i.e. rock or peat), it directs the tidal flow and plays an important role in the allocation of channels and shoals.

The current study aims to investigate the impact of the geometry on the morphodynamic development in a tidal basin by application of a process-based, numerical model. We performed multiple model runs with the geometry of the Western Scheldt starting from a flat bed and continuing for 200 years. The generated channel-shoal patterns are visually comparable to observed patterns in the Western Scheldt tidal basin. Even after 200 years of calculations the model results show ongoing developments, albeit at an increasingly slower rate.

No significant differences were observed by introducing salt-fresh water density differences or variations in sediment grain size and the morphological factor. Factors that contributed significantly to a better model performance were adaptations in the bed slope factor, inclusion of overtides at the seaward boundary, 3D instead of 2D calculations and, in some cases, the inclusion of non-erodible layers, dredging and dumping activities and higher river discharge. Once calibrated with different bed slope factor values, good results were obtained with both the van Rijn and Engelund-Hansen sediment transport formulations.

The generated hypsometries are comparable to the measured hypsometry and all cases show a better fit with the measured hypsometry over time. An analysis based on an adapted BSS shows an increasing skill within the first decades. This is attributed to the imposed geometry inducing initial channel-shoal development that fit the measured patterns. On the longer term (> 50 years), the BSS declines in most cases leading to negative values. A possible cause is the ongoing pattern development, but also the reallocation of sediment along the basin and sediment export from the basin may play a significant role. The latter processes are governed by the overtides introduced at the boundary or generated internally in the model domain.

The current study shows that a fixed geometry and schematized tidal forcing can significantly explain morphological patterns observed in the Western Scheldt tidal embayment, both in terms of size as in terms of allocation of the patterns. This suggests that the geometry of a tidal basin plays a significant role in the basins' morphological characteristics and probably also in its further morphodynamic development.

6 Process-based, morphodynamic hindcast of decadal deposition patterns in San Pablo Bay 1856-1887[5]

Abstract

Process-based numerical models are able to describe morphodynamic developments over time. The results of these models depend on model parameter settings related to sediment transport, bed composition or the schematization of boundary conditions. However, the (historical) parameter settings are often not known in detail.

The objective of the current research is to assess the value of process-based morphodynamic modeling by hindcasting decadal morphodynamic development in San Pablo Bay, California, USA. Focus is on the 1856-1887 period in which upstream hydraulic mining resulted in a high sediment input to the Bay. The Delft3D model includes processes like wind waves, salt and fresh water interactions and graded sediment transport for both sand and mud fractions. Model outcomes are evaluated against measured bathymetric developments and include an extensive sensitivity analysis on model parameter settings..

The results show that a model applying best-guess model parameter settings can predict decadal morphodynamic developments in San Pablo Bay quite well. Results are consistent for variations of parameter settings. Variations in sediment concentration, river discharge and wind waves have the most effect on deposition volumes, whereas waves have the most impact on sediment allocation within San Pablo Bay. Brier Skill Scores have values around 0.25 with a maximum of 0.43 (qualified as 'good') although higher values (up to 0.65) were obtained in sub-areas of San Pablo Bay. The results can be improved by systematically adjusting model parameters, although this would be a computationally demanding task. Including extreme events may further enhance model performance.

---

[5] A slightly adapted version of this chapter was submitetd to Journal of Geophysical Research - Earth Surface Processes by authors Mick van der Wegen, Bruce Jaffe and Dano Roelvink

6.1 Introduction

6.1.1 Scope

Adequate management of an estuarine system requires understanding of the processes governing bathymetric evolution. Examples of relevant morphodynamic developments are the evolution of salt marshes [Allen (2000), Temmerman et al. (2005)], the siltation of access channels to ports [Kirby (2002)], and erosion exposing legacy contaminants [Van Geen and Luoma, (1999), Higgins et al. (2007)].

The estuarine morphodynamic system is characterized by interaction of different spatial scales and time scales. This implies that bed level changes at a particular location and time are a function of the larger morphodynamic system and should not be studied separately. Furthermore, changes in bed level can be considerable on a long time scale. This implies that significant morphodynamic developments may take place that are not recognized or understood over relatively short time spans of years.

Bathymetric surveys on the decadal time scale are invaluable for developing insight into the morphodynamic behavior of an estuarine system. These measurements are, however, costly and require discipline in timing and coverage of the area of interest. Detailed data are scarce and often even non-existant, especially when longer time scales (> 50 years) are considered. Furthermore, data analysis by itself may not be able to distinguish explicitly dominant forcing mechanisms and processes.

Numerical, process-based modeling efforts may enhance the understanding of the morphodynamic estuarine system. Given a well-calibrated hydrodynamic model, sediment transport and bathymetric change can be calculated to make hindcasts and long-term predictions of the estuarine bed level changes. Also, potential impacts of anthropogenic works like groyne construction, land reclamations or potential impacts of climate change and sea level rise can be investigated with such a model.

The value of morphodynamic models depends to a high degree on empirical formulations, such as sediment transport formulae or roughness formulations. These empirical equations include constants associated with confidence intervals that are determined by laboratory tests or, more seldom, in nature itself. For example, critical shear stress for the initiation of sediment motion depends on the character of the bed material, its compaction rate and even on the (seasonally varying) presence of biomass. Thus it becomes a function of time. Furthermore, characteristic values for critical shear stress will vary for deeper layers under the bed surface and across the area of interest.

One can argue that for an accurate model prediction of sedimentation and erosion patterns it is necessary to measure these bed characteristics across the whole model domain and over the whole time span of interest. This would be a demanding, if not impossible, task both in terms of time and finance and it can be questioned whether it is really necessary to do so. Parameter estimation can also be based on inverse data assimilation techniques. For example, Yang and Hamrick (2003) explored the possibility of deriving cohesive sediment transport parameter values based on iteratively comparing model outcomes with adjusted parameter settings to an extensive dataset. This methodology, however, would require extensive data for assimilation.

Another approach, applied in the current research, is by simply assuming 'reasonable and averaged values' of modeling parameters. This method is in particular useful if data are limited or lacking, for example in case of historical hindcasts. Sensitivity analysis on parameter variation would subsequently indicate estimated ranges of the value of the model outcomes. Thus, insight can be gained on dominant forcing mechanisms in the system and sensitivity to changing boundary conditions, such as, for example, changing river discharge regime and sea level rise.

Different attempts have been made to study morphodynamic development using process-based numerical models. For idealized tidal conditions, Seminara and Tubino (2001), Schramkowski et al. (2002) and Van Leeuwen and De Swart (2004) (amongst others) show morphodynamic modeling efforts investigating 2D pattern formation characteristics. On an embayment length scale Hibma et al. (2003a). Friedrichs and Aubrey (1996), Schuttelaars and De Swart (2000) and Lanzoni and Seminara (2002) (amongst others) adopt a schematized 1D approach to investigate equilibrium conditions of a longitudinal embayment profile. Hibma et al (2003b,c), Van der Wegen and Roelvink (2008) and Van der Wegen et al. (2008) show that process-based, 2D numerical morphodynamic models lead to relatively stable patterns and bed profiles in about 100 km long embayments under idealized circumstances like constant forcing conditions and uniform sand distribution over the modeling domain. Van der Wegen and Roelvink (2008) and Van der Wegen et al. (2008) distinguish a time scale of pattern formation (~decades) and a time scale related to the development of the full embayment (~ millennia). Others applied a similar methodology to describe decadal development and pattern formation in more realistic geometries related to Waddenzee inlets, the Netherlands [Marciano (2005), Dissanayake (2007) and Dastgheib et al. (2008)] or the Western Scheldt, the Netherlands [Van der Wegen and Roelvink (2008)].

The studies mentioned above are governed by tide dominated sand transports. In a study closely related to this research Ganju et al. (in press) focused on a mud dominated estuary with large freshwater flow. They modeled decadal morphodynamic development in Suisun Bay, a sub-embayment of San Francisco Bay which is located adjacent to San Pablo Bay. Ganju et al. (in press) found an average error of 37% for bathymetric change over individual depth ranges and poor spatial amplitude correlation performance on a model cell-by-cell, though spatial phase correlation was better, with61% of the domain correctly indicated as erosional or depositional. The present study differs from the Ganju et al (in press) study in the sense that an advanced estimate is made on the initial bed composition and that emphasis is put on the impact of variations of model parameter settings.

6.1.2   Aim of the study

The objective of the current research is to assess the ability of process-based, morphodynamic modeling to hindcast decadal morphodynamic development in an existing estuary with a complex sand-mud environment.

6.1.3   Approach

The model is being validated against an extensive and relatively long (~150 years) dataset from San Francisco Estuary describing sedimentation patterns of a major sediment pulse due to hydraulic mining mid 19th century. This research focuses on a (measured) depositional period (1856-1887) in San Pablo Bay which is a sub-embayment of the San Francisco Estuary (Jaffe et al., 2007). The reason why this period was chosen is that it describes the impact of a major forcing signal, i.e. temporal and excessive sediment supply by two major rivers.

Since historical data on model boundary conditions and bed composition are lacking, particular emphasis is put on a sensitivity analysis of forcing mechanisms such as river discharge and sediment concentration and variations in sediment transport parameters.

The following sections describe, in order, the historical context of the morphodynamic developments in San Pablo Bay, the model setup and the model parameter settings, model results in comparison to measured developments and finally model results in terms of an extensive sensitivity analysis.

6.2 Description of San Francisco Estuary

Extensive literature is available on the San Francisco Estuary. The short introduction given below only aims at highlighting the most relevant aspects for the current research and is based on the review by Kimmerer (2004), if not otherwise stated.

6.2.1 Geometry

San Francisco Estuary is a drowned tectonically reshaped river valley. It consists of a number of interconnected sub-embayments (*Figure 6.1*). Two main rivers, the Sacramento River and the San Joaquin River, meet in the area referred to as the Delta, which consists of a complex network of channels, sloughs and shallow lakes. The river flow discharges from the Delta to Suisun Bay, via the relatively narrow Carquinez Strait, to San Pablo Bay and Central Bay and, finally, through the narrow and deep Golden Gate into the sea. This area is referred to as the northern reach. Additionally, Central Bay connects to South Bay extending southward. This latter bay, however, receives considerably less river flow.

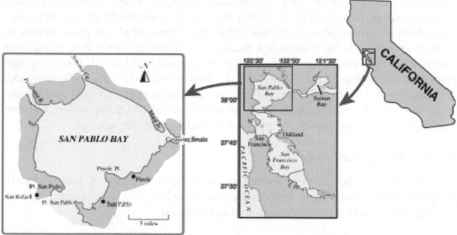

Figure 6.1 Location of San Pablo Bay

In Suisun Bay the main channel from the Delta splits into two main channels, the Northern passage and Southern passage, of which the latter is the main shipping channel and has a depth of about 10-15 m. The channels are sandy and silty with occasional mud banks. The shallows (0-6 m below MSL) between the main channels and north of the Suisun Bay are mud covered and partly intertidal. Carquinez Strait has an alluvial bed with a maximum depth of about 35 m below MSL and is flanked by rock.

The bathymetry of San Pablo Bay is characterized by a single main channel, about 20-30 m deep, that connects Carquinez Strait to San Pablo Strait. The latter has a maximum depth of about 40 m. The main channel has a silty bed with sandy and muddy patches [Locke (1971)]. Two other, much smaller and shallower, channels in the northern part of San Pablo Bay are dredge-maintained. The extensive shallow areas both south and north of the main channel are muddy (with particles largely smaller than 4 μm) and cover about 80 % of the Bay. About 90% of these shallows is less than 4 m deep and silty intertidal flats are present at the coastline edges [Locke (1971)].

Central Bay is relatively deep and sandy and comprises a much lower percentage of shallow, mud covered area than San Pablo Bay, Suisun Bay or South Bay. Central Bay is covered with a range of bed forms depending on the sand grain size and local tidal velocities [Rubin and Mc Culloch (1979)].

6.2.2 Hydrodynamics

The river discharge regime is characterized by a relatively dry season (summer/autumn) and a wet season (winter/spring). However, amounts and distribution of yearly discharge over time are highly variable. Peak discharges may amount to 17,800 m^3/s, but can be 300 m^3/s during dry years, whereas low discharges during autumn may not exceed 100 m^3/s. The San Joaquin River is responsible for about 10-15 % of the discharge and the Sacramento River for about 80 %. The remaining discharge originates from minor tributaries. During high discharges the Yolo Bypass, a major managed floodplain of the Sacramento River, may convey up to 60 % of the Sacramento River discharge. Values and timing of river discharges are subject to human interferences by means of managed reservoir releases in the watershed and water export from the Delta for the purpose of fresh water supply to Southern California.

The tide near the Golden Gate is highly irregular [Smith (1980)]. It consists of a dominant semidiurnal M2 component, a spring-neap tidal variation and even considerable longer period fluctuations. Median tidal range is about 1.8 m. In the estuary itself friction and reflection processes complicate the tidal behavior. Water levels and velocities in North Bay are in phase (i.e. a progressive wave) and water level amplitudes slowly decay, whereas in South Bay the tidal wave shows resonant behavior so that a time lag between water levels and velocities develops and the water level amplitude amplifies near the head. During high river discharge no tidal influence is present at Sacramento (155 km upstream from the Golden Gate), whereas minor tidal fluctuations are observed during low river discharges. Water levels at Golden Gate show a slight increase (~ cm) with high river flows.

The tidally averaged mean salt intrusion up the estuary depends primarily on fresh water flow and to a lesser extent on spring-neap tidal variations. The 2 psu isohaline is found most often in Suisun Bay and dam water release during dry periods is managed in such a way that that salt does not intrude farther landward. High river flows may cause salinities to decrease to less than 5 psu east of San Pablo Bay. Stratification and gravitational circulation occurs in particular during neap tides, when tidally driven mixing processes are weak. The location varies depending on the river discharge. Ganju and Schoellhamer (2006) give an example of stratification measured near Benicia Bridge in Carquinez Strait. Low river flows allow for salinities of about 25 psu near Benicia Bridge and considerably reduce stratification.

6.2.3   Sediment dynamics

It is generally assumed that most sediment to the estuary is supplied by the two major rivers and that tributaries discharging directly into San Pablo Bay have a minor role in sediment supply [Porterfield (1980), Krone (1979)], although Mc Kee et al. (2006)] suggest that sediment supply by tributaries will become more important as sediment load from the two major rivers declines.   Based on measurements, Wright and Schoellhamer (2004) and references therein suggest that suspended load dominates bed load by approximately an order of magnitude. Import from the sea, if any, is considered hardly relevant, although no measurements confirmed this yet.

Hydraulic mining from 1850 to1884 caused an excess supply of sediments mainly to the Sacramento River [Gilbert (1917)]. Krone (1979) suggests that large volumes of this sediment settled in the Delta, Suisun Bay and San Pablo Bay during the period of hydraulic mining. Measurements and comparison of historical bathymetries confirm this process. As hydraulic mining stopped Suisun Bay started to erode [Cappiella et al. (1999)]. San Pablo Bay only became erosional mid 20th century [Jaffe et al. (1998, 2007)]. Hydraulic works like reservoir construction in the upstream watershed probably enhanced the erosion process by upstream trapping of sediments. Based on data from 1957 to 2001 in the Sacramento River Wright and Schoellhamer (2004) indicate a decreasing trend in suspended sediment discharge for a given flow and attribute this to depletion of erodible sediments from hydraulic mining, trapping of sediment in reservoirs, altered land use and construction of protection works and levees.

Figure 6.8 (a,c,d) show measured 1856 and 1887 bathymetries as well as the deposition and erosion patterns of this period. Jaffe et al. (1998) describe how these figures were composed.

6.3 Model set up

Use is made of a process-based, 3D, numerical model (Delft3D). Delft 3D solves the shallow water equations, including the k-ε turbulence closure model, and applies a horizontal curvilinear grid with sigma layers for vertical grid resolution.   It allows for salt-fresh water density variations, separate formulae for mud transport and sand transport, and variations in bed composition and specification (for example, bed layers with different percentages of mud and sand and spatial variation of critical shear stress). The impact of wind and waves (SWAN) is added, so that, for example, the effects of wind set up and increased shear stress due to waves in shallow water are taken into account. Lesser et al. (2004) and section 1.4 describe the model in more detail.

The model parameter settings for the San Pablo Bay case study are described in detail in the following sections. Partly, the settings are simplified to reduce the computational time. Partly, model parameters are simplified since detailed data in time and space are simply not available.

6.3.1   Model domain

The model domain comprises the area from Rio Vista and Antioch in the Delta (in two separate branches for the Sacramento River and the San Joaquin River) to Richmond in Central Bay and thus includes San Pablo Bay as well as Carquinez Strait and Suisun Bay, see *Figure 6.1* and *Figure 6.2*. This domain is large enough to have a negligible boundary definition effect on the area of interest and the domain is small enough to allow for relatively fast runs (~ 36 hours for about 4 hydrodynamically modeled months on a 'decent' PC (3.0 Ghz, 3.25 Gb RAM, quad core)).

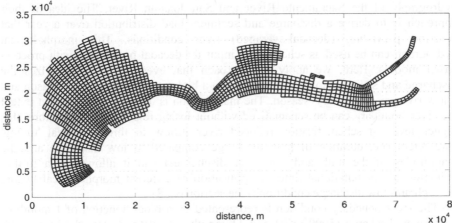

Figure 6.2 Numerical grid of model covering San Pablo Bay and Suisun Bay. The upper branch at the landward side represents the Sacramento River and the lower branch the San Joaquin River.

A curvi-linear grid is applied on the domain and the condition for a stable and accurate computation (Courant number < 10) is met with a grid cell size of approximately 100 by 150 m and a time step of 2 minutes. Density currents and wave effects require a 3D approach. 15 Sigma layers describe the vertical grid distribution, which is somewhat finer near the bed and the water surface to adequately resolve velocity shear near the bed and wind and wave effects.

The bathymetry is composed of measured bathymetries of Central Bay, San Pablo Bay and Suisun Bay measured in the years 1855, 1856 and 1867 respectively. The measured bathymetries were corrected for the fact that their datum is a spatially varying MLLW and the models' datum is NAVD 88, which is constant over the modeling domain.

6.3.2   Hydrodynamic boundary conditions

Historical data describing boundary conditions for the 1856-1887 period are not available. An additional complication for a proper setting of the boundary definition is that the system is characterized by considerable inter-annual and seasonal variations of both the river discharge and the tide. However, the current research aims at discriminating major forcing mechanisms in the system which may very well be covered by a high level of boundary condition schematization. Another reason for input schematization is the high computational effort that is involved in hindcasting

30 years of morphodynamic development. A 'schematized' hydrodynamic year would capture the high variation of the system and reduce computation efforts. The following sections describe this input reduction.

Peak sediment loads are dominant in the sediment supply to the San Francisco Estuary. For example, based on more recent data McKee et al. (2006) suggest that the January 1997 flood transported about 11 percent of the total 9-year load, and that almost 10% of the yearly sediment load can be delivered in one day, and over 40% within seven days for an extremely wet year. This suggests that the morphodynamic system is strongly characterized by a short wet period and a longer dry period.

Ganju et al (2008) describe a methodology to derive a 'yearly morphological hydrograph' of the Sacramento River and San Joaquin River. The idea of their approach is to derive a discharge and sediment load distribution over a year that describe prevailing (decadal-averaged) river conditions. This morphological hydrograph can be used as schematized input for decadal predictions with process-based models. Here, we apply an approach that even further schematizes the discharge and sediment load distribution into a high river discharge 'wet' season and a low river discharge 'dry' season. The main reason is to investigate to what extent the river boundary can be schematized without losing relevance. Furthermore, this higher level of schematization reduced river inflow to three practical 'tuning' parameters (i.e. duration of high discharge compared to low discharge and the magnitudes of the high and low river discharges) and it allows for a further reduction of computational time by application of different morphological factors for high and low discharge conditions (see section 6.3.8).

The river boundary condition is represented by a block function of 1 month of high river discharge (5000 m^3/s) and 11 months of relatively low discharge (350 m^3/s), of which 80% is assigned to the Sacramento River branch and 20% to the San Joaquin branch. At first glance the wet season discharge might seem large, but it should reflect a discharge that is representative for the sediment pulse, which relates stronger than linearly to water discharge. These river boundary conditions are input for a hydrodynamic model describing a larger model domain ranging from the Delta to 20 km offshore including all sub-embayments of San Francisco Estuary (Elias et al. in prep), which was run on the historical bathymetry again composed of bathymetries measured closest to 1856. The tidal forcing of this model consists of a representative monthly tidal cycle of 10 constituents. Running this representative cycle for 12 months describes similar characteristics as a full year of tides including tidal constituents with timescale effects larger than a month. Elias et al. in prep describe this input reduction technique in more detail.

Discharge time series were derived from the large model at the location of the current models' landward boundary conditions and water levels were derived at the current model's seaward boundary. Only the signal of major tidal constituents leading to a long-term average transport, i.e. M_2 M_4, O_1 and K_1 at the boundary locations were selected. Of these, the combination of M2 and M4 clearly leads to a tidal asymmetry and associated net transport (e.g. Van de Kreeke and Robaczewska, 1993). Less well-known, the combination of M2, O1 and K1 has a similar effect, since the frequencies of O1 and K1 exactly add up to that of M2 (e.g. Hoitink et al., 2003). These latter constituents still describe a neap-spring tidal cycle, which would complicate the schematization of a wet month especially in combination with the morphodynamic updating technique described in the next section. For example, the sediment pulse starting from the landward boundary could always reach San Pablo

Bay at a spring tide and yearly variations would thus be disregarded. A final step in input reduction was made by combining O_1 and K_1 into one artificial diurnal component C_1 with frequency half of that of M2, so that the neap-spring tidal cycle vanishes from the input. Leading principle in this reduction is that the sediment transport generated by the C_1 signal over a month is equal to the transport generated by the O_1 and K_1 components. Lesser (2009) describes the methodology in more detail. In this procedure the effects of N2 and S2 components have been neglected; these are expected to lead to variations within a spring-neap cycle but not to important long-term average effects. Non-tidal forcings at the seaward boundary condition, such as El Nino effects, storm surges are neglected as well.

In accordance with general observations (Kimmerer et al 2004) salt concentration is set constant at zero at the landward boundary and at 25 psu at the seaward boundary.

6.3.3   Wind and waves

Prevailing wind conditions are schematized by a diurnal sinusoidal signal varying from 0 at midnight to 7 m/s at noon uniformly distributed over the domain based on the wind climatology described by Hayes et al. (1984) and additional analysis of wind data from http://wwwcimis.water.ca.gov/cimis/welcome.jsp. For the wet period and 6 months of dry period the wind comes from the west and for the remaining 5 months of dry period from the south east mimicking the seasonal variations in wind field.

Every hour the SWAN model uses wind and hydrodynamic data from the flow calculation to generate a wave field and returns resulting adapted hydrodynamic parameters to the flow module. A detailed description of the SWAN wave model and its application in Delft 3D can be found respectively at the SWAN homepage (http://vlm089.citg.tudelft.nl/swan/index.htm) and Lesser et al. (2004).

6.3.4   Sediment transport

Sediment transport is modeled by a 3D advection-diffusion equation including expressions for erosion and deposition as source terms. Fall velocities and formulations for erosion rate and deposition rate depend on the sediment. The transport of cohesive mud is modeled by the Partheniades-Krone formulations (Partheniades, 1965).

$$E = MS_e(\tau_{cw}, \tau_{cr,e})$$
$$D = w_s c_b S_d(\tau_{cw}, \tau_{cr,d})$$
(6.1)

In which

E   erosion flux [kg/m2/s]
M   erosion parameter [kg/m2/s]
D   deposition flux [kg/m2/s]
w_s sediment fall velocity [m/s]
c_b near bottom concentration [kg/m^3]
τ_{cw} maximum shear stress due to waves and current [N/m^2]
$\tau_{cr,e}$ critical shear stress for erosion [N/m^2]
$\tau_{cr,d}$ critical shear stress for deposition [N/m^2]

and

$$S_e(\tau_{cw}, \tau_{cr,e}) = \left(\frac{\tau_{cw}}{\tau_{cr,e}} - 1\right) \; for \; \tau_{cw} > \tau_{cr,e}$$

$$= 0 \qquad\qquad for \; \tau_{cw} \le \tau_{cr,e}$$

(6.2)

$$S_d(\tau_{cw}, \tau_{cr,d}) = \left(1 - \frac{\tau_{cw}}{\tau_{cr,d}}\right) \; for \; \tau_{cw} < \tau_{cr,d}$$

$$= 0 \qquad\qquad for \; \tau_{cw} \ge \tau_{cr,d}$$

(6.3)

For the transport of non-cohesive sediment, Van Rijn's (2000) approach is followed. Lesser et al (2004) describe the implementation of the transport formulations in Delft 3D.

6.3.5 Sediment fractions

Data on bed composition in San Pablo Bay are very limited. Only Locke (1971) presents an overview of bed composition in 1968 (measured in December) and 1969 and 1970 (both measured in February). Bottom sediments of San Pablo Bay are primarily clay and silt, except for the main channel, which is sand in places covered with mud banks. However, as specified by equations (6.1)-(6.3), the numerical model requires input in terms of erosion factor, critical erosion/deposition shear stress and sediment fall velocity.

Preliminary model runs show that reasonable results can be obtained only by applying multiple sediment fractions (i.e. both sand in the channels and mud on the shoals). Also, applying different fractions allows studying the behavior of different fractions at the same time. The selection of the fractions is based on adding a range around limited data values and 'best guesses'.

For the sandy fractions 1, 2 and 3 diameters of 500, 300 and 150 μm were chosen following characteristic values by Locke (1971).

Table 4 shows various mud transport parameters and their measured values and values applied in process-based models from literature. Values applied in the current research are given

Table 5. In addition, for all fractions, the erosion parameter (M) is $2.0*10^{-4}$ $kg/m^2/s$, the bulk density is 1200 kg/m^3, the dry bed density is 500 kg/m^3 and, following suggestions by Winterwerp et al. (2004, pp 144-148), the critical shear stress for deposition ($T_{d,cr}$) is set at 1000 N/m^2. This implies that deposition takes place continuously and is not limited by a critical shear stress value above which no deposition takes place.

	τ_e (N/m^2)	M (kg/m^2/s)	w *10^{-3} (m/s)	Location
Teeter (1987)	0.1-0.4	1.3-4.7x10^5		Alcatraz disposal site
Sternberg et al. (1986)	0.049		2	Shoal near Richmond
Kranck and Milligan (1992)			1-10	near ATF location in San Pablo Bay
Jones (2008) [1]	0.26-2.56 0.13-0.92 0.12-2.08			ATF location San Pablo Bay
McDonald and Cheng (1997) [2]	0.3-0.4	1-5x10^{-5}	0.4-1	North Bays Modeling study
Ganju and Schoellhamer (2007) [3]	0.1/1.0- 0.11/1.1	2-1.8x10^{-5}	0.1/0.25- 0.11/0.275	Suisun Bay Modeling study

Table 4 Mud transport parameters from literature. [1] Depth dependent data from bed level to approx. 27 cm into the bed; [2] McDonald and Cheng (1997) used critical shear stress as calibration parameter for fitting to SSC; [3] Ganju and Schoellhamer (2007) use two different mud fractions

	$T_{e,cr}$ (N/m^2)	w (mm/s)
Mud 1- m1	0.8	0.4
Mud 2- m2	0.5	0.28
Mud 3- m3	0.3	0.16
Mud 4- m4	0.1	0.064
Mud 5- m5	0.05	0.024

Table 5 Mud specifications applied in the current model.

6.3.6   Initial bed composition

Assuming just muddy shallows and sandy channels could be wrong and lead to significant errors in the model. Van der Wegen et al (in prep) describe a methodology to generate a bed composition of different sediment fractions by using a process-based numerical modeling approach similar to the current research. They use a bed composition model that applies the active layer concept by Hirano (1971).

The model of Van der Wegen et al (in prep.) starts from a uniform distribution of 8 sediment fractions. Each time step water levels, velocities and sediment transports are calculated (see also next section). However, the bed level itself is kept constant during the calculation so that the divergence of the sediment field only leads to a redistribution of the sediment fractions over the domain bathymetry. For example, due to high prevailing shear stresses, mud is washed out from the channels to the shoals so that the channels become sandier and the shoal muddier. During a run of 4 months Van der Wegen et al initially observe a large reallocation of sediments which vanishes after a number of days. Following, smaller and ongoing developments are attributed to the autonomous development of the system.

The initial bed composition of the current model (presented in Figure 6.6) is generated according to this methodology under low discharge conditions. The final bed composition of the active layer (the upper 20 cm) from the 'bed composition run' forms the initial bed composition of the current model and it is assumed that this composition prevails over the entire 8 m of sediments available in the bed. As expected, the bed composition run results in sandy deeper seaward parts of the channel due to the large shear stresses. The heavier mud fractions are clearly present in the shallower and landward portions of the main channel, whereas the lighter mud fractions are distributed more on the shallows. The lightest mud fraction is hardly present, which means that this fraction was washed out since it could not withstand the prevailing hydraulic conditions. M2 and m3 fractions dominate the mud presence.

6.3.7   Sediment concentration boundary conditions

The model requires sediment concentrations at the landward boundary. For this, we use the method Ganju et al. (2008) developed a method to estimate daily sediment loads after the start of hydraulic mining 150 years ago from average annual river discharge. The method uses historical rainfall data in Sacramento (from 1 October 1850 onwards) and unimpaired flow estimates (from 1906 onwards) and current data as a proxy for the historical data. They predicted sediment loads based on the estimated historical discharges via the relationship suggested by Muller and Forstner (1968):

$$Q_s = aQ_r^{\,b+1} \tag{6.4}$$

In which

Q_s annual sediment load, kg/s
Q_r annual river discharge, m³/s
a, b   calibration coefficients

Parameter b represents the erosive power of the stream, which is a function of stream/floodplain morphology. Parameter a is related to sediment availability. For the current watershed this parameter varies strongly with time, in accordance to hydraulic mining, urbanization, and retention of sediment behind dams. Based on a comparison to decadal sediment loads estimated by Gilbert (1917) for the period 1849-1914 and Porterfield (1980) for the period 1909-1966, Ganju et al. (2008) suggest a constant value of b=0.13 and a time varying value of a=0.02 (before hydraulic mining) to 0.13 (when hydraulic mining stopped in 1884) and slowly decreasing to approximately 0.03 for more recent decades. For the current research sediment concentrations at the landward boundary were based on b=0.13 and a= 0.13 for the 1856-1887 period. In case of a river discharge of 5,000 m³/s this leads to a concentration of about 390 mg/l. In the model each of the mud fraction concentrations is set at 300 mg/l. Although the total sum of concentrations is considerably larger than the suggested value, it allows studying the individual behavior of the 5 fractions with a similar concentration as the one suggested by Ganju et al. (2008). The sand fractions are attributed concentrations of 10 % of the mud fractions.

At the seaward side no sediment concentration is prescribed, although suspended sediment, advected by tidal movement, is present. This means that sediment would leave the model domain and ideally come back after turning of the tide through the model boundary. To include this process, the model applies a Thatcher-Harleman time lag at the boundaries that stores sediment concentrations and reintroduces these at the boundary with a time lag of 120 minutes.

6.3.8   *Morphodynamic updating*

For the model to run efficiently, a morphological factor (MF) multiplies every time step the bed level changes calculated from the divergence of the sediment transport field. This approach requires that (upscaled) bed level variations within a tidal cycle are small compared to the water depth, so that bed level changes have negligible influence on the hydrodynamics. Details on this methodology can be found in Roelvink (2006) and Van der Wegen and Roelvink (2008).

During the wet period a MF of 30 is applied. Sensitivity analysis showed that during the dry season a MF value of 82.5 could be applied. The reason that a larger MF can be used for the dry period is that preliminary model runs showed that transports during the dry period appeared typically up to 2 orders of magnitudes smaller than during the wet period (i.e. in the 1856-1887 deposition period). The sensitivity analysis in later sections will show the difference with smaller values of the morphological factor.

In summary, in order to reproduce 30 years of morphological development, the model calculates first 1 month of high river discharge with a morphological factor of 30 with western wind, then 2 months of low river discharge with a MF of 82.5 and western wind and finally 2 months of low river discharge with a MF of 82.5 and south-eastern wind.

A problem initially encountered during the runs was unreasonably high sedimentation rates in the eastern portion of both river branches in the east during the dry season. Under conditions of high sediment concentration inflow and smaller local shear stresses due to the absence of a high river discharge, m1 and m2 fractions deposited in the river branches even to such an extent that the branches were blocked and bed levels at the river boundaries became dry (above high water) and no river discharge could enter the Bay anymore. The problem is attributed to the fact that the Delta is highly schematized in the model. In reality the Delta is a more complex network and would allow sediments to settle in areas not directly impacting the river flow. Besides, one of the reasons why gold mining was stopped in the 19th century was related to the increasing problem that people encountered in keeping the rivers navigable by dredging. The model would thus correctly reproduce, at least qualitatively, the observed historical sedimentation. The problem was solved by application of the 'dredging' option, maintaining the river branches at a water depth of 2 m during the dry period. Any surplus material was removed from the model domain during the run. This did not significantly affect the model results in San Pablo Bay, since only little sediment reaches San Pablo Bay from the Delta area during the dry period.

6.3.9   *Description sensitivity analysis*

As described in previous sections the model applies a number of simplifications in the parameter settings partly to reduce the model run times by input reduction techniques and partly because detailed data (both in time and in space) do not exist. In order to address the impact of likely model parameter variations an extensive sensitivity analysis is carried out by systematically varying these parameters (one by one) and comparing the outcomes to a 'standard case'. This implies that, apart from the 'standard' run, 22 sensitivity runs were carried out. The variations can be roughly subdivided into 'mud characteristics' and 'hydrodynamics' and are described in Table 6.

	Standard	Low	High
Mud characteristics			
Concentrations (mg/l)	300	250	350
Critical shear stress (N/m^2)	0.05-0.8	0.04-0.64	0.06-0.96
Erosion coefficient (kg/m^2/s)	$1*10^{-4}$	$5*10^{-5}$	$2*10^{-4}$
Sediment fall velocity (mm/s)	0.03-0.5	0.024-0.4	0.036-0.6
Horizontal eddy diffusivity (m^2/s)	1	0.1	10
Hydrodynamics			
Manning roughness	0.2	0.17	0.23
River discharge (m^3/s)	5,000	4,000	6,000
Tidal amplitude (m)		-10%	+10%
Wind (m/s)	7	5	9
Waves	Yes	No	
Salt/density effects	Yes	No	
Dimension	3D	2D	
Morphological factor	Wet: 30 Dry: 82.5	Wet: 3 Dry: 8.25	

Table 6 Model parameter specification and parameter variations for sensitivity analysis.

6.4 Results

In order to address the main objective of the study, the modeled bathymetric evolution is compared to measured data using deposition volumes and a Brier Skill Score (BSS) (e.g. Murphy and Epstein (1989), Sutherland et al, 2004, van Rijn et al, 2003). The influence of variations in model parameter settings on the model results is also investigated. Because the focus of the current study is on morphodynamic aspects, the discussion of results emphasizes sediment transport and morphology. If not otherwise stated, the presented results reflect the standard case. First, the model results are described generally.

6.4.1 Model performance

All models produced regular tidal variations in water level and velocity after a one-day spin-up interval. During the wet month salt does not intrude farther than mid San Pablo Bay, whereas salt intrudes up to east Suisun Bay during the dry months, which is largely in accordance with general observations (Monismith et al., 2002). Maximum waves by westerly winds occur in the south eastern part of San Pablo Bay near the entrance to Carquinez Strait and do not exceed 0.45 m in height and a period of 2.5 s. South-eastern winds cause waves with a maximum height of 0.35

and a maximum period of 2 s on the north western shoals. Mean water level is about 20 cm higher during the wet than during the dry.

During the wet month the high sediment transport causes a deposition pulse through the model domain, starting in Suisun Bay and arriving some days later in San Pablo Bay. This pulse can be observed in the development of cumulative mud transports (*Figure 6.4*a, see *Figure 6.3* for definition of locations). The m5 fraction is not shown in this figure since it does not play a major role in the deposition patterns. Sand transports are not shown in Figure 4 because they are typically at least an order of magnitude smaller than the mud transports.

During the wet period the inflow through cross-section CS (import) exceeds the outflow through cross-section PSB (export) for all mud fractions (*Figure 6.4*a and *Figure 6.5*a,b). The finer the mud fraction, the faster the imported and exported amounts become (almost) equal. When this happens net deposition in San Pablo Bay does not take place anymore. The probable reason why this occurs is that the initial deposition leads to a bathymetry that enhances the process of sediment by-passing through San Pablo Bay. A shallower bathymetry or a more confined channel area lead to higher velocities so that more sediment is kept into suspension which is subsequently washed out seaward by tide residual flows (of which the river discharge will be the main component).

The equilibrium between import en export is not reached for the heavier mud fractions in the modeled wet period (*Figure 6.4*a), although it might be reached on longer time scales. *Figure 6.5*c shows that the deposit is primarily composed of the m3 fraction. The finer the mud fraction, the less is deposited as percentage of the inflow (*Figure 6.5*d).

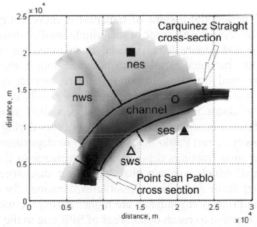

Figure 6.3 Definition of San Pablo Bay sub-areas on 1856 measured bathymetry. Darker shades are deeper. NWS- North West Shoal; NES - North East Shoal; SWS- South West Shoal; SES- South East Shoal; Area symbols are used in following figures.

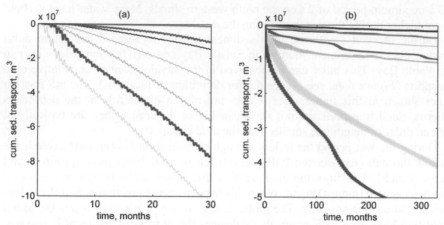

Figure 6.4 Cumulative mud transports through Carquinez Strait (CS) cross-section (light lines) and Point San Pablo (PSB) cross-section (dark lines). Locations of cross sections shown in Figure 6.3. Negative values indicate seaward transport. Transports and time are multiplied by the morphological factor. Black denotes m1 fraction; Green denotes m2 fraction; Red denotes m3 fraction; Blue denotes m4 fraction; (a) During 30months 'wet' conditions. Mind that black and green dark lines (m1-CS and m2-PSB transport) almost coincide with origin; (b) During 330 months of 'dry' conditions. Mind that light black and dark green lines (m1-CS and m2-PSB transport) almost coincide;

During the 'dry' period the pattern of transport is more complex. Cumulative transports are typically almost one order of magnitude smaller than during the 'wet' period (*Figure 6.4*a,b and *Figure 6.5*a,b). Tidal fluctuations are clearly present, in particular for the finer fractions. Equilibrium between import and export develops faster for the heavier mud fractions. In contrast to the heavier mud fractions, the export of the finest fraction exceeds the import (*Figure 6.4*b, c). Furthermore, the eroded amount of the finest fractions exceeds the deposited amount of the other larger fractions, so that net erosion of San Pablo Bay occurs during the dry. Without the fine fractions, however, San Pablo Bay would still be depositional during the dry period. For the larger mud fractions deposition as percentage of the inflow is lower during the dry period, which suggests that high river discharge during the wet creates conditions that favor deposition. Possible explanations for this behavior are that concentrations and transport gradients are higher during the wet season, that the larger mud fractions are able to reach other areas of SPB due to the river flow or that tidal velocity fluctuations are more damped.

McKee et al., (2006) and Ganju and Schoellhamer, (2006) recently observed prevailing landward transport of sediments towards Suisun Bay during low river discharge conditions. They attributed this transport to strong gravitational circulation in Carquinez Strait in combination with high suspended-sediment concentrations in San Pablo Bay. This seems in contrast with the results of the current study. A number of explanations can be given. First of all the current model may not adequately represent the process of gravitational circulation in Carquinez Strait. This process is strongly sensitive to the settings of the κ-ε turbulence model and would require a separate validation (see for example Ganju and Schoellhamer 2007), which is outside the scope of the current research. The modeled gravitational circulation may also show sensitivity to the highly schematized tidal conditions (excluding extremes in the neap spring tidal cycle) and river discharge (excluding variations during the year). Secondly, the landward transport trend entering Suisun Bay through Carquinez Strait depends on the SSC. Ganju and Schoellhamer (2007) suggest that sediment transport entering Suisun Bay during flood is related to the high SSC induced by wind wave suspension in San Pablo Bay. This high SCC is prescribed as boundary condition in their model leading to the observed landward transport during low discharge conditions. However, apart from the wind waves in San Pablo Bay, complex conditions in Carquinez Strait itself might also be responsible for the high SSC.

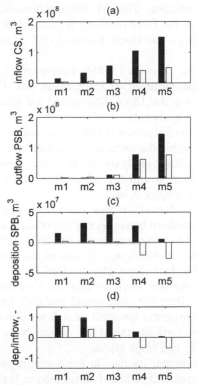

Figure 6.5 Transport and deposition of mud fractions during wet and dry periods: (a) inflow of mud through Carquinez Strait cross-section; (b) outflow of mud through Point San Pablo cross-section; (c) deposition of mud in San Pablo Bay; (d) ratio of deposition to inflow; Black bars indicate the wet period and white bars indicate the dry period. Transports are multiplied by the morphological factor. Erosion shown as negative deposition. Deposition and erosion volumes include porosity.

Carquinez Strait is relatively deep compared to Suisun Bay and the channel in San Pablo Bay so that it may act as a sediment trap during high river discharge. Our model confirms this (not shown). Sediment may then be re-suspended during low river discharge. Further, Carquinez Strait geometry is irregular and strongly bended which probably leads local sediment trapping and considerable secondary currents ('spiral flow'), Finally, Ganju and Schoellhamer (2005) describe the presence of an estuarine turbidity maximum in Carquinez Strait. These processes may not be captured by the current relatively coarse grid and would require a more detailed model of the area. Despite the fact that the current model does not lead to landward transport through Carquinez Strait, transports during dry conditions are low compared to sediment transport during wet conditions so that it is reasonable to assume that major transport mechanisms are covered by the model.

Figure 6.6 and Figure 6.7 show the bed sediment composition of the upper, active layer (20 cm below the bed surface) at the start and at the end of the standard model run. The main differences are that the extent of the sand decreases as the channel narrows, which was observed, and that the sides of the main channel and the NES shallows become muddier. The m1 fraction deposits along the sides of the channel and deposition of the finer fractions is concentrated in the shallows at the margins of San Pablo Bay. The finest fractions (m4 and m5) are almost entirely washed out.

Figure 6.8 shows the measured 1856 bathymetry (a), the measured 1887 bathymetry (c) and the modeled1887bathymetry (e). Cumulative erosion and sedimentation patterns over the 1856-1887 period are given in (d) and (f) for the measured and modeled bathymetries.

Figure 6.8 (b) shows the difference (modeled -/- measured) of these patterns.

Modeled patterns resemble the order of magnitude and roughly the allocation of the measured volumes. Closer analysis of the model results indicates that major deposits near CS cross-section and the edges along the main channel are mainly caused by the m1 fraction (Figure 6.7d).

The large quantity of deposition that occurred along the sides of the main channel (at least for the southern banks) is not reproduced well by the model (Figure 8). The presence of m1 and s3 fractions seems to play a crucial role (compare Figure 6.6 c, d to Figure 6.7c, d). Furthermore, it should be taken into account that this area is subject to considerable bed slope effects. Bed slope effects for mud are not included in the model and bed slope effects for sand are highly parameterized. Reducing the bed slope effect for sand would probably improve the results in this region. It would also diminish the modeled sand deposition in the centre of the main channel near Pinole Point which might be caused by sand sloping down from southern channel banks by bed slope effects.

Minor modeled depositional patches in the northern part of San Pablo Bay are not present in the measurements, which is attributed to the fact that tidal exchange and discharges from Petulama River and Sonoma Creek (see *Figure 6.1*) are not included in the model schematization.

Closer analysis of tidal (diurnal) residual transports during the wet and dry periods reveals the dominant mechanisms in sediment re-allocation. The analysis is made after the wet period when the characteristics are most pronounced (Figure 6.9). The wet period tidal residual transports (Figure 6.9a) are typically an order of magnitude larger than the dry period tidal residual transports (Figure 9b). During the wet period the largest gradients in the tidal residual transports are found in the main channel near Carquinez Strait and Point San Pablo. These are the locations where, respectively, the most sediment deposits and erodes. Residual transports on the shallows are negligible.

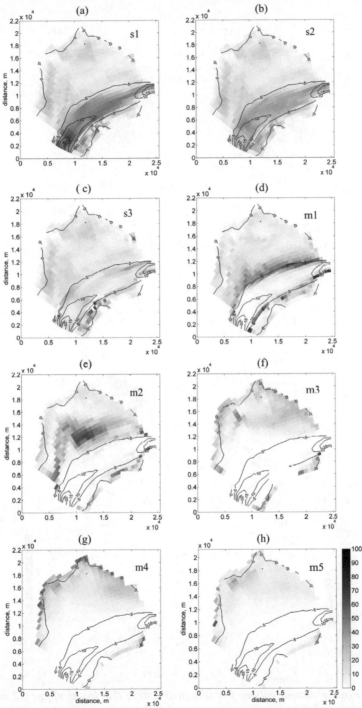

Figure 6.6 *Initial sediment volume fraction distribution (%) in the upper 0.2 m of the bay floor for (a) s1 fraction; (b) s2 fraction; (c) s3 fraction; (d) m1 fraction; (e) m2 fraction; (f) m3 fraction; (g) m4 fraction; (h) m5 fraction. The 5m contour lines show the 1856 bed level.*

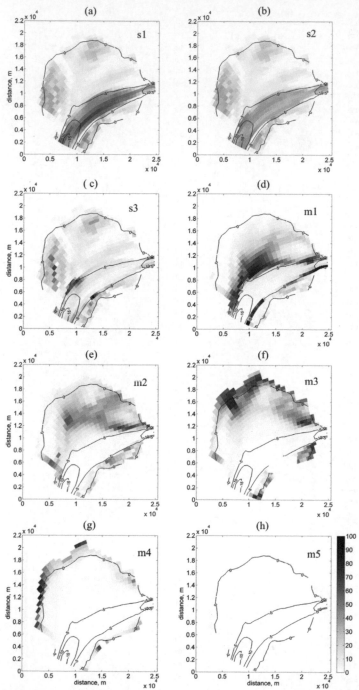

Figure 6.7 *Final sediment volume fraction distribution (%) in the upper 0.2 m of the bay floor for (a) s1 fraction; (b) s2 fraction; (c) s3 fraction; (d) m1 fraction; (e) m2 fraction; (f) m3 fraction; (g) m4 fraction; (h) m5 fraction. The 5m contour lines show the modeled 1887 bed level.*

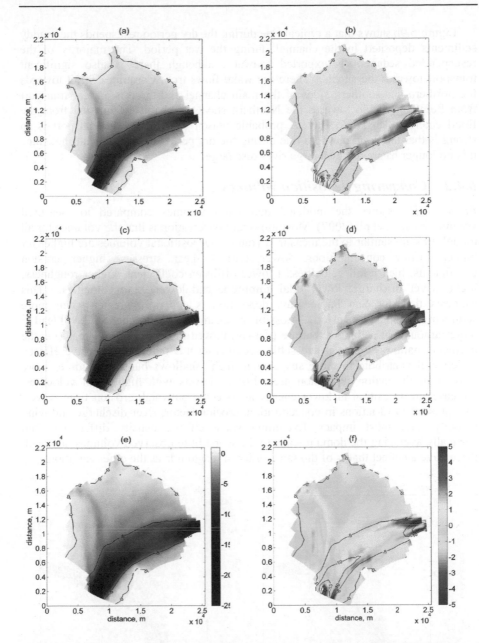

Figure 6.8 (a) Measured 1887 bathymetry with 5 m contour lines; (b) Measured erosion (cold colors) and sedimentation (warm colors) patterns in 1856-1887 period with 1856 5 m contour lines; (c) Modeled 1887 bathymetry with 5 m contour lines; (d) Modeled erosion (cold colors) and sedimentation (warm colors) patterns in 1856-1887 period with 1856 5 m contour lines; (e) Measured 1856 bathymetry with 5 m contour lines; (f) Difference in erosion and sedimentation for patterns of measured and modeled 1856-1887 period with 1856 5 m contour lines. Warm colors indicate higher modeled bed levels. Cold colors indicate higher measured bed levels;

Figure 6.9b shows that a typical tide during the dry period resuspends the muddy sediments deposited in the channel during the wet period. The majority of the resuspended sediment is exported seaward, although there is also significant transport towards the shoals. There are wake flows from Carquinez Strait towards the northern and southern sides of the main channel and there is a clear transport from Point San Pablo towards the Northern shoals, probably originating from the flood entering San Pablo Bay. A probable reason why this flood transport is so strong is that the flood during a tide during the dry period is relatively strong, since it is no longer hampered by a large river discharge.

6.4.2   Comparing deposition volumes

*Figure 6.10*a shows the modeled deposition volumes compared to measured volumes by Jaffe et al. (2007). Most important observation is that the volumes for all model runs are similar to the measured volume. Depositional volumes are higher for higher inflow concentrations, lower critical shear stresses, higher erosion coefficients, higher fall velocities, a lower diffusion coefficient, higher roughness, higher river discharge, lower tidal amplitude and lower wind velocities. This suggests that, for San Pablo Bay, deposition is generally greater when more sediment is kept in suspension. A possible mechanism is that higher SSC will lead to larger amounts of sediment that are (slowly) moved to lower energy areas (i.e. the northern shallows) where it can settle. Countering this tendency are wave effects, which will contribute to more suspension in the shallows but also leads to more erosion in the major deposition area. For sediments with higher fall velocities, increases in deposition in lower energy areas could outweigh erosion from waves. The prescribed variations in concentration, erosion factor, river discharge and wind velocity have most impact. Excluding wave effects, density differences and vertically averaging the domain leads to the worst fit to observed volumes, although these have an effect that is of the same order of magnitude as the other variations.

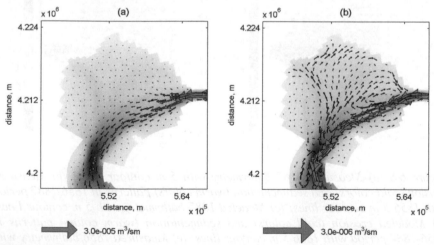

Figure 6.9 Tide residual sediment transport (m³/sm) at the end of the wet period. Conditions (a) during wet period; (b) during dry period. Note different scales for transport vectors.

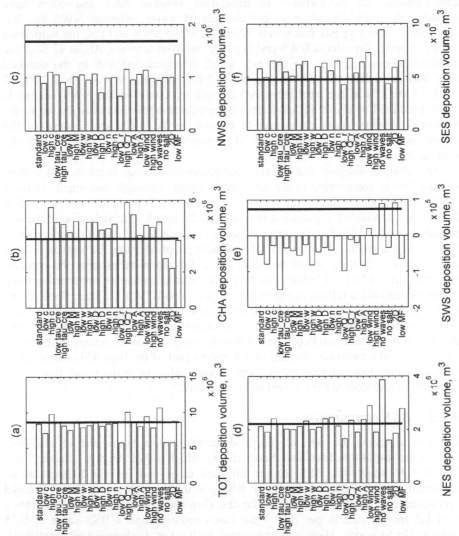

Figure 6.10 Sensitivity analysis results of deposition volumes in San Pablo Bay including porosity. Solid line denotes measured volume after Jaffe et al (2007); (a) total deposition, (b) Channel deposition, (c) NWS deposition, (d) NES deposition, (e) SWS deposition, (f) SES deposition.

An interesting observation is that minimum and maximum parameter values do not always lead to smaller or larger deposition volumes compared to the standard case. The cases with a low and high sediment fall velocity, for example, both lead to smaller deposition volumes. This is attributed to San Pablo Bay geometry. Changing model parameters will not only change deposition and erosion, but also the locations where these processes take place. For example, in shallow areas the impact of a higher sediment fall velocity may be considerably different than in the deep channel.

The impact of changing parameter settings on sediment allocation within San Pablo Bay may be considerable. The absence of waves, for example, leads to approximately 35 % increase in deposition volume. NES and SWS have considerably more deposition in the absence of waves, whereas NWS has less deposition. This suggests that waves erode NES and SWS and that the suspended sediment partly deposits at NWS and is partly removed seaward. About 47 % of the measured deposition is in the channel, which is reproduced well by the standard case. Measured deposition volumes in NWS, NES, SWS and SES are about 21%, 25 %, 1 % and 6% of the total measured deposition. The modeled NWS volume is under predicted by about 40%, whereas the modeled NES volume is similar to the measured volume. Most cases predict erosion in the SWS area instead of the measured deposition. The modeled SES deposition slightly overestimates the measured volumes.

This analysis suggests that the standard parameter settings result in a deposition volume that is comparable to the measured volume; although, too much sediment settles in the main channel and too little sediment deposits in the western part of San Pablo Bay. A possible explanation is related to the rough schematization of the wet period. In reality, floods take place in peaks of a number of days instead of being spread over a month. The amount of sediment supplied might be similar in both cases, but a peak flood would transport the sediments more towards the western, seaward parts of San Pablo Bay. Another process responsible for transporting more sediment seaward is related to the timing of peak floods. When a peak flood occurs after a period in which considerable deposition already took place, the sediment will also be moved more seaward. Another possible explanation for underestimating deposition in western San Pablo Bay is that the early deposition results in shallower conditions and a narrower channel in the eastern part of the Bay. This would cause high velocities that enhance erosion and hinder sediment settling and, thus, more transport of sediment to the western part of the Bay.

6.4.3   BSS comparison

In order to assess the skill of morphodynamic models Sutherland et al (2004) suggest the use of the Brier Skill Score (BSS), which is explained in Chapter 1.4.5. *Figure 6.11* shows the BSS for the different cases and for the different areas. Maximum BSS is 0.31 and maximum BSS_{vR} is 0.47, which means that the model performance is 'good' for that particular case (low roughness). The BSS_{vR} assumes a δ of 0.2 m³ (or: 0.2 m per m².) Most cases score around a BSS value of 0.25 (reasonable to good). High concentrations, high river discharge, high amplitude, excluding waves and the 2D approach lead to the worst results. NES has the highest BSS of 0.64 (excellent) NWS scores comparable to the total area and the channel area scores generally slightly lower values. Scores for SWS are generally bad and scores for SES are slightly lower than the full domain scores. By far the weakest BSS values for all areas except NWS are found for the case with no waves. Waves thus have a significant effect on sediment allocation in San Pablo Bay.

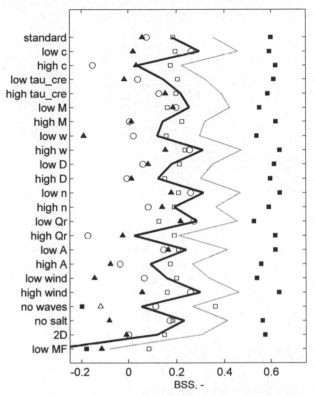

Figure 6.11 BSS-values. 'o' denotes BSS for full San Pablo Bay domain; '□' denotes BSS for full San Pablo Bay domain including correction for measurement accuracy by equation (1.15); Thick solid line denotes channel; small open squares denote NWS; small filled squares denote NES; open triangles denote SWS (largely smaller than -0.5); filled triangles denote SES.

These observations are also roughly reflected in the scores of α and β (*Figure 6.12*a,b). Best α and β scores are for NWS, NES, and the channel, whereas SWS and SES have the worst scores. Phasing of the prediction could be improved on the southern shoals. Including the amplitude performance with β shows much more variance in the different cases than considering phases only with α, which suggests that the phasing of the pattern is less sensitive to the changing parameter settings than the pattern amplitudes. Logically, γ values, shown in *Figure 6.12*c, follow the deposition volume performance (Figure 10), although they do not discriminate between under or over prediction. It is noted that volume underestimates result in relatively high γ values compared to similar magnitude volume overestimates.

The SWS area scores worst and the NES scores best both in terms of the deposition volume and the BSS. However, on a more detailed level volumetric (or (γ) values) and BSS performances are not similar. A low sediment fall velocity, for example, leads to a similar deposition volume, but the BSS of this case is one of the weakest. Another example is that the low river discharge leads to a higher BSS than the case of the high river discharge, although it scores worse in terms of deposition volume.

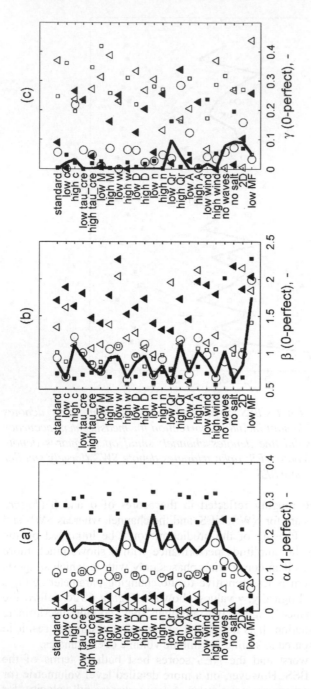

Figure 6.12 BSS-parameter values. 'o' denotes BSS for full San Pablo Bay domain; Thick solid line denotes channel; small open squares denote NWS; small filled squares denote NES; small open triangles denote SWS; ; small filled triangles denote SES. (a) 'α' is a measure of phase error; (b) 'β' is a measure of both the amplitude and phase error; (c) 'γ' is a measure of the mean error.

6.4.4 Model improvement

The aim of the study is not to come to optimal model predictions, but rather to investigate the impact of model parameter variations on model performance. Nevertheless, it is worth exploring how the results can be improved. The previous section suggests that strategies to improve the model that adjust parameter settings are not easily deduced from the model results. One reason is that variations in a particular model parameter have different effects on the volumetric change and the BSS. Another reason is that an increase in a model parameter (for example the sediment fall velocity) does not necessarily lead to an expected result (increase in deposition), because the increase also changes the allocation of sediment. A third reason is that it is possible that changing model parameters would lead to a different sensitivity to changes of other model parameters.

Model performance improvement is an iterative process, in which parameters are changed one at a time and sensitivity analyses are carried out for every model parameter variation. It is even possible that the order in which the model parameters are varied would lead to different results. Despite potential problems, an attempt is made to increase model performance by adapting model parameter according to their performance in the earlier analysis. Five additional runs were made that are specified in

Table 7. The case with higher river discharge for the additional runs included a 10 day wet period (instead of 1 month) with a river discharge of 8000 m3/s. All cases excluded m4 and m5 fractions at the inflow boundary, since these would wash out completely. Results are presented in

Figure *6.13* and show that case 004 leads to best BSS results with a overall score of 0.43, although deposition volumes are underestimated by about 15%. Results of case 001 being worse than the standard case confirm that improving model outcomes is not automatically reached by combining best performing cases of previous runs, but that it would rather include a (time demanding) process of parameter optimization.

A final remark is made on extreme conditions. The present study applies highly averaged forcing conditions, such as the wind field and the boundary conditions. It is very well possible that extreme conditions have a significant impact on the sediment allocation. An example is the possible effects of El Nino on the water level variations at sea, the river discharge or local wind conditions. The present study shows that applying average conditions already leads to good qualitative and quantitative results. Even better results may be obtained by investigating the impacts of extreme conditions. This is, however, considered outside the scope of the present study.

Run	low c	high w	low n	low Q_r	Higher Q_r	Low bed slope effect	high wind
001		*		*			*
002	*		*				*
003	*	*	*	*			*
004	*		*			*	*
005	*		*		*	*	*

Table 7 Adapted model settings compared to standard case

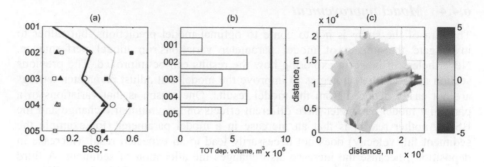

Figure 6.13 Model results for improved parameter settings (a) BSS; (b) Total deposition volumes; (c) cumulative erosion and sedimentation (m) for case 004.

6.5 Summary and conclusions

Morphodynamic developments in San Pablo Bay between 1856 and 1887 were hindcast with a process-based, numerical model. The model is 3D and uses the shallow water equations, a k-ε turbulence model, salt water (density) gradients, 3 sand fractions and 5 mud fractions, as well as a scheme that enhances morphodynamic developments. The model was simplified partly to reduce the computational time and partly due to the fact that (historic) data on sediment characteristics were lacking. The boundary conditions are highly schematized by describing only the M_2, M_4, and C_1 tidal components at the seaward boundary and a block function describing 1 month of high river discharge (5000 m^3/s) and 11 months of low river discharge (350 m^3/s) at the landward boundary. Also assumptions on inflow sediment concentrations and mud characteristics needed to be made. Morphodynamic developments were enhanced by multiplying the bed level changes by a factor of 30 for the wet period and a factor of 82.5 for the dry period.

When compared to measured morphodynamic developments over 30 years, the investigation shows that the model is capable of adequately reproducing decadal deposition volumes in San Pablo Bay. The allocation of the sediments within the Bay is good as indicated by the BSS, which has a maximum of 0.43.

The model results show distinct sediment transport patterns for high and low river discharges. Part of the sediment is deposited in the main channel and the rest conveyed seaward during high river discharges, whereas the sediment deposited in the main channel are transported towards the northern shoals (about 10-20%) and seaward (about 80-90%) during periods of low river discharge. Relatively fine mud fractions are washed completely out of the Bay. Sand transports are typically an order of magnitude smaller than mud transports.

The model results are based on estimated model parameter settings. Sensitivity analysis on these settings yields results that similar to a standard case. This holds for realistic variations in mud characteristics, like critical shear stress or fall velocity, as well as in hydrodynamic characteristics like diffusion coefficient or tidal amplitude. Including physical processes like waves and density differences as well as a 3D approach are necessary to obtain reasonable results. Sediment concentration, river discharge and waves have the most effect on deposition volumes, whereas waves have the most impact on sediment allocation within San Pablo Bay.

The best strategy for adjusting parameter settings to improve model performance is not easily deduced from the model results. Preliminary efforts show that this would be a time demanding task of iteratively varying model parameters once at a time to find the optimal settings in the model parameter domain.

Model performance may be further improved by systematically adjusting the model parameter settings. However, the best strategy to do this is not easily deduced from the model results. Preliminary efforts show that this would be a time demanding task of iteratively varying model parameters once at a time to find the optimal settings in the model parameter domain. Another way to improve model results may be by including the impact of extreme events such as local storms, occasionally high river discharges or El Nino effects.

6.6 Acknowledgements

The research is part of the USGS CASCaDE climate change project (CASCaDE contribution 17). The authors want to acknowledge USGS Priority Ecosystem Studies, CALFED as well as the UNESCO-IHE research fund for making this research financially possible.

7 Conclusions and recommendations

7.1 Conclusions

7.1.1 Introduction

The previous chapters include detailed descriptions and analyses of different modeling studies and can be read as separate studies. They are related to each other in the sense that subsequent chapters cover smaller time scales and increasing complexity in model formulation. Similarly they range from more fundamental studies without the possibility of proper data validation to studies where validation against data plays a crucial role in the assessment of the model performance. All chapters are related to the main aim of this study *to investigate the governing processes and characteristics that drive morphodynamic evolution in alluvial estuaries by application of a process-based numerical model.*

This section focuses on answering the research questions posed in the introduction. These questions were formulated to specify more clearly the research aim. The detailed conclusions of the different chapters will not be repeated here. However, in answering the research questions reference is made to the relevant sections in this thesis.

7.1.2 Answer to research questions

1. *How can morphodynamic 'equilibrium' be defined?*

The models described in the current research did not reach a state of equilibrium in the strict sense. Tide residual transports and morphodynamic evolution remain continuously present during the simulations despite the constant forcing conditions. Nevertheless, model results showed a decay in (the rates of) energy dissipation and tide residual transports (sections 3.3.2 and 3.3.3).

However, a number of parameters were distinguished that can act as indicators for the morphodynamic stability of tidal basins. These are ebb/flood duration (which should be equal), phase difference between water levels and velocities (which should be 90^0 out of phase to prevent Stokes drift), velocity and water level amplitudes (which should be similar for ebb and flood and constant along the basin) and energy dissipation (which should be similar during ebb and flood and constant along the basin). This list of indicators is not necessarily sufficient, since obeying the indicators' criteria for equilibrium, the basins still show significant tide residual transport (sections 2.4.4 and 3.3.2). The presence of tidal constituents higher than M_2 at the boundary or internally generated in the model domain probably accounts for a significant effect on the morphodynamic developments in the basin (section 2.4.4).

The best indicator for morphodynamic equilibrium would be the spatial tide residual transport gradient. When this gradient is low and spatially constant, one may even observe a relationship between tidal prism (P) and cross-sectional area (A) which is constant along the basin and seemingly constant over decades. This modeled PA relationship corresponds with an empirically derived PA relationship. However, at the same time, the small tide residual transport gradients cause adaptation (in the current research: deepening) of the basin and an increase of cross-sectional area for an equal tidal prism (non-equilibrium). This process leads to a PA relationship with changing coefficients (section 4.3.2).

2. *What timescales can be distinguished in long-term, 2D process-based morphodynamic modeling?*

A first characteristic time scale is related to bed level developments. When starting from a flat or linearly sloping bed model results show that channel-shoal patterns evolve quite rapidly (section 2.3.3). Within decades small channels combine and merge into larger channels so that finally a network develops that shows a much smaller evolution rate. Characteristic length scales of the channel-shoal pattern are primarily determined by an adaptation length scale of the flow and an adaptation length scale of the bed topography (sections 2.3.4 and 2.4.2).

A second timescale is related to the development of the (width averaged) depth profile along the basin. Main parameters involved in the evolution of this profile are the tidal constituents and in particular the development of overtides. The time scale of this process depends on the length of the basin and appears to be (generally) an order of magnitude smaller than the time scale of the pattern development. It is noted that, after their initially rapid growth, the channel-shoal patterns adapt to the changing conditions of the longitudinal profile with a similar, slow time scale (section 2.4.1).

The development of the longitudinal profile is also apparent in the evolution of the tidal prism (P) and cross-sectional area (A) relationship along a tidal basin (sections 2.3.7 and chapter 4). At any moment in time the PA relationship is constant along the basin and equilibrium seems to be present on a decadal time scale. This is probably caused by the fact that cross-sections that deviate too much are rapidly smoothened with adjacent cross-sections in the basin. However, model results on long (~century) time scales show slow, but continuously increasing cross-sections for similar tidal prisms. This tidal channel evolution is related to (very) small but persisting spatial gradients in tide residual transport. The time scale of the cross-sectional evolution is comparable to the time scale of the longitudinal profile mentioned in the previous section.

A third time scale is related to (excess) sediment supply. Model results show that the magnitude and rate of energy dissipation (or: shear stresses and sediment transports) decrease over longer time scales (section 3.3.3). Under extreme conditions, however, these parameters may show a short period of increase. An example is the formation of a channel-shoal pattern that develops from an initially flat bed. The channel-shoal pattern develops due to an instability that is inherent to the non-linear system of the shallow water equations, sediment transports and bed level development. This causes water to flow through relatively narrow channels with relatively high velocities and results in higher energy dissipation compared to flat, uniform cross-sections (section 3.3.3). Another example of a process that leads to increasing energy dissipation is the sediment supply from banks under the bank erosion algorithm applied in the current research (sections 1.4.3.4 and 3.3.3). A final example is the excess sediment supply to San Pablo Bay by hydraulic mining in the period 1856-1887. This supply resulted in high deposition in San Pablo Bay. The deposition led to a shallower bathymetry that enhanced sediment transports (or: energy dissipation) and the seaward sediment conveyance capacity through San Pablo Bay (6.4.1). The sediment supply time scale is related to the sediment supply itself, but also to the capacity of the whole system to deal with such an amount of sediments. The bank erosion led to a gradual and continuous supply of sediments to the basin. The impact of this supply was of the order of millennia in an 80 km long basin, but would have been shorter when the initial condition of the basin geometry were deeper and wider. First decades of the hydraulic mining San Pablo Bay could not deal with the amount of sediments provided. However, when velocities increased due to the fact that San Pablo Bay became shallower, more sediment was conveyed through San Pablo Bay and deposition rates decreased. This effect was significant within decades.

A final time scale worth mentioning within the framework of the current work is the time scale of sea level rise and land subsidence processes. The long term simulations for 80 km long basins require millennia before reaching a state of 'near-equilibrium', especially when bank erosion is taken into account. This intersects with the time scale of sea level rise, which has been relatively constant last 3000 years after a period of rapid sea level rise. Alluvial estuarine systems might still be in a phase of (significant) adaptation to these relatively constant conditions of sea level, especially when the systems are 'saturated' with sediments due to excess supply or due to the fact that the systems are relatively young due to recent sea level rise induced flooding of alluvial coastal plains.

3.   *What processes are relevant in long-term morphodynamic evolution?*

The current work shows the value of both highly schematized model setups and modeling approaches that include more detailed process descriptions. It appears that, regarding predictions on longer time scales and larger domains, the number of processes and parameters that seem relevant can be decreased (compare Chapter 2 to Chapter 6). This is partly due to the simple fact that hardly any long-term empirical data are available so that including these data in the model (i.e. boundary conditions, bed composition) just does not make sense. Partly this can be attributed to simpler long-term system behavior. For example, the seasonal development of a particular bed forms and flocculation processes might be of interest to predict accurate and instantaneous sediment transports. However, seasonal fluctuations in roughness and floc size seem to become irrelevant for decadal morphodynamic predictions so that these can be incorporated in mean values of the Manning coefficient and, for example, a constant fall velocity. After an extensive sensitivity analysis model results suggest that even applying best-guess values for sediment characteristics lead to qualitatively good results in terms of both decadal volumetric changes and pattern development in San Pablo Bay (Chapter 6).

4.   *Does long-term morphodynamic modeling under free and idealized*
     conditions lead to a preferred geometry?

In a fixed narrow basin morphological features develop that resemble an alternating bar pattern. The bends of the channel that meanders along these alternate bars are located at the banks and become quite deep (sections 2.3.3 and 0). This suggests that considerable erosion would take place if the banks were erodible. Once a simple bank erosion algorithm is introduced, the main erosion of the banks indeed takes place in the outer banks of the channel. Visual observations suggest that the basin widening process is directly related to the meandering amplitude of the main channel in the basin (section 0).

The geometry that develops describes a landward linearly decaying cross-section, which is different from exponentially decaying geometries often referred to in literature. Although development continues even after millennia, the linear profile is continuously present during this evolution (section 0). The linear decay is similar to the equilibrium profile of the geometry of a short basin. Although tidal characteristics in a long basin (including for example resonant behavior) are quite different from short basin conditions, these appear not to play a significant role in the shape of the geometry.

5. *What is the influence of geometry on the morphodynamic evolution of an alluvial estuary?*

Allowing for erodible banks leads to free geometry development with a landward linearly decaying cross-section. In a basin with fixed banks, a channel-shoal pattern develops that is restricted by the banks. The (very) deep outer bends of the main meandering channel are an indicator of this (sections 2.3.3 and 0). The width to depth ratio of a fixed banks basin compared to an erodible banks basin is therefore larger. For a given tidal prism fixed banks with deep and narrow cross-sections lead to smaller cross-sectional areas than erodible banks with wide and shallow cross-sections. This is attributed to smaller effects of friction in relatively deep and narrow channels (section 0).

The Western Scheldt geometry has a significant effect on its observed channel-shoal pattern (Chapter 5). Starting from a flat bed and imposing the major tidal forcing which is observed at the Western Scheldt mouth, initial patterns forced by the Western Scheldt geometry develop in correlation to the observed bathymetry. Best agreement (Brier Skill Score) is observed after decades, after which the BSS of most cases declines, due to sediment export from the basin. Some cases show improved BSS on longer time scales (centuries). This is attributed to relocation of sediments along the basin by tidal constituents higher than M_2.

High Brier Skill Scores were obtained in hindcasting decadal deposition and erosion patterns in San Pablo Bay, California (section 6.4.3). The 3D model developed included a full range of interacting processes like turbulence, salt and fresh water, multiple sand and mud fractions, wind and wind waves. Rough schematization of wind conditions and river discharges and applying reasonable values for sediment characteristics led to good results in terms of both volumetric changes and pattern development. Sensitivity analysis showed that model results were not fundamentally and qualitatively different after variations of the process parameter settings. This suggests that both the bathymetric state itself and the geometry of the Bay are dominant factors in the morphodynamic evolution of San Pablo Bay (Chapter 6).

7.2 Recommendations

1. The morphological factor

The morphological factor, as it is applied in the current work, is a very useful tool in long-term morphodynamic modeling. It leads generally to stable computations and considerable time saving. However, good sensitivity analysis is required to set the value of the factor. Apart from instabilities in the computation the only guideline in its application is that bed level changes should typically not exceed 5% of the water depth. A strict rule on the height of the morphological factor is not available. Further research is required on a stricter guideline and its application in cases with graded sediments and bed composition in which armoring process play a role. Another case of interest is the role of the morphological factor in the interaction between sediments and biological entities, which have their own time scale and may be subject to, for example, seasonal fluctuations.

2. Model validation

As process-based models are able to predict morphodynamic developments on a decadal time scale, there is an increasing need for tools that are able to assess and value the results and to compare them adequately with observed data. The current research shows examples of comparison in terms of empirical equilibrium relationships, volumes, hypsometry and visual patterns and the Brier Skill Score, each of these having a limitation. Also these parameters may not capture system characteristics such as channel and network structures or bed level slopes characteristics. Better pattern recognition algorithms that are able to classify and compare model results with observations, will help to assess the shortcomings and improve the capacity of morphodynamic modeling.

3. Open source software

The Delft3D software provides a state-of-the-art, long-term morphodynamic modeling tool. One of its major limitations is that it is (to date) not open source software. From a scientific point of view this is not desirable. Access to (super-) computer facilities is restricted and further, free development of the code (and thus knowledge) is suppressed. In the end, this will lead to other software systems incorporating more and better techniques and a loss of scientific interest for further development of the Delft3D software. It is therefore recommended to develop an open source Delft3D version for scientific research.

References

Ahnert, F. (1960), Estuarine meanders in the Chesapeake Bay area, Geographical Review 50, 390-401.

Airy, G.B., (1845), Tides and waves, Encyc. Metrop., Art. 192.

ASCE Task committee on hydraulics, bank mechanics and modeling of river width adjustment, (1998a), River width adjustment. I: processes and mechanisms, in Journal of Hydraulic Engineering, Vol. 124, No. 9, pp. 881-902.

ASCE Task committee on hydraulics, bank mechanics and modeling of river width adjustment, (1998b), River width adjustment. II: Modeling, Journal of Hydraulic Engineering, Vol. 124, No. 9, pp. 903-917.

Aubrey, D.G., and Speer, P.E. 1985. A Study of Non-linear Tidal Propagation in shallow Inlet/Estuarine Systems Part I: Observations. Estuarine. Coastal and Shelf Science, 185-205

Bagnold, R.A. (1966), An approach to the sediment transport problem, General Physics Geological Survey, Prof. paper 422-I, Washington.

Bates, P.D., and J.M. Hervouet (1999), A new method for moving boundary hydrodynamic problems in shallow water, Proc. R. Soc. London, Ser. A, 455, 3107-3128.

Beets, D.J. and A.J.F. van der Spek (2000), The holocene evolution of the barrier and the back-barrier basins of Belgium and the Netherlands as a function of late Weichselian morphology, relative sea-level rise and sediment supply, Netherlands Journal of Geosciences, 79 (1), 3-16.

Blott, S.J., K.Pye, D. Van der Wal and A. Neal (2006), Long-term morphological change and its causes in the Mersey Estuary, NW England, Geomorphology, 81, 185-206.

Booij, N., R.C. Ris and L.H. Holthuijsen, 1999. A third-generation wave model for coastal regions, Part I, Model description and validation, *J.Geoph.Research*, 104, C4, 7649-7666)

Boon, J.D., (1975), Tidal discharge asymmetry in a salt marsh drainage system, Limnol. Oceonagr., Vol. 20, 71-80.

Boon, J.D.III and R.J. Byrne (1981), On basin hypsometry and the morphodynamic response of coastal inlet systems, Marine Geology, 40, 27-48.

Borsje, B.W., De Vries, M.B., Hulscher, S.J.M.H., De Boer, G.J., 2008. Modeling large-scale cohesive sediment transport with the inclusion of small-scale biological activity. Estuarine, Coastal and Shelf Science 78, 468-480. *doi: 10.1016 / j.ecss.2008.01.009*Brown, E. I., 1928. Inlets on sandy coasts. Proc. Am. Soc. Civ. Eng., LIV, pp. 505–553.

Bruun, P., 1978. Stability of tidal inlets, Theory and engineering, Elsevier, Scientific Publishing Company, Amsterdam-Oxford-New York.

Burau, J. R., Gartner, J. W., and Stacey, M. T., 1998. Results from the hydrodynamic element of the 1994 entrapment zone study in Suisun Bay, in: Kimmerer, W. ed., Report of the 1994 entrapment zone study. Technical Report 56, Interagency Ecological Program for the San Francisco Bay/Delta Estuary, pp 13- 55.

Cappiella, K., Malzone, C. , Smith R., Jaffe, B.E., 1999. Sedimentation and bathymetric change in Suisun Bay 1876-1990. USGS Open File Report, 9-563.

Cayocca, F., 2001. Long-term morphological modeling of a tidal inlet: the Arcachon Basin, France. Coastal Engineering 42, 115-142.

Coevelt, E.M., A. Hibma and M.J.F Stive (2003) Feedback mechanisms in channel-shoal formation, in: Proceedings of the fifth International Symposium on Coastal Engineering and Science of Coastal Sediment Processes, Coastal Sediments '03, ASCE, Clearwater Beach, Florida, USA, (available at CD-Rom published by World Scientific Publishing) ISBN 981-238-422-7

Conomos, T.J., Peterson, D.H., 1977. Suspended-particle transport and circulation in San Francisco Bay: an overview. In: Wiley, M. (Ed.), Estuarine Processes, vol. 2. Academic Press, New York, pp. 82–97.

D'Alpaos, A., S. Lanzoni and M. Marani (2005), Tidal network ontogeny: Channel initiation and early development, Journal of Geophysical Research, Vol.110, F02001, doi: 10.1029/2004JF000182.

D'Alpaos, A., S. Lanzoni, M. Marani and A. Rinaldo (2007), Landscape ecolution in tidal embayments: Modeling the interplay of erosion, sedimentation, and vegetation dynamics, Journal of Geophysical Research, Vol.112, F01008, doi:10.1029/2006JF000537.

D'Alpaos, A., S. Lanzoni, M. Marani, and A. Rinaldo (2009), On the O'Brien-Jarrett-Marchi law, Rend.Lincei, 20, 225-236, doi:101007/s12210-009-0052-x.

D'Alpaos, A., S. Lanzoni, M. Marani, and A. Rinaldo (2010), On the tidal prism-channel area relations, J.Geophys. Res., 115, F01003, doi:10.1029/2008JF001243.

Dalrymple, R.W. and R.N. Rhodes (1995), Estuarine dunes and bars. In:, (ed.), Geomorphology and Sedimentology of Estuaries, edited by G.M.E. Perillo, Developments in Sedimentology 53, Elsevier Science, New York, pp 359-422.

Dastgheib, A., Roelvink, J.A., Wang, Z.B. (2008), Long-term process-based morphological modeling of the Marsdiep Tidal Basin, Marine Geology, 256, pp. 90–100.

Defina, A. (2000), Two-dimensional shallow water flow equations for partly dry areas, Water Resources Research, 36, 3251-3264.

DEFRA (2005), Review and formalisation of geomorphological concepts and approaches for estuaries, Department for Environment, Food and Rural Affairs/Environment Agency, Report No: FD2116/TR2, Lincoln, United Kingdom.

De Jong, H., Gerritsen, F., 1984. Stability parameters of Western Scheldt estuary. In: B.L. Edge (ed.), Proceedings of the 19th Coastal Engineering Conference, ASCE, Vol.3, p.3078-3093, 1984

De Swart , H. Zimmerman (accepted), Morphodynamics of Tidal Inlet Systems, Annu. Rev. Fluid Mech. 2009. 41:X—X, doi: 10.1146/annurev.fluid.010908.165159.

De Vriend, H.J., Zyserman, J., Nicholson, J., Roelvink, J.A.., Pechon, P., Southgate,. H.N., 1993a. Medium-term 2DH coastal area modeling. Coastal Engineering 21 (1-3), 193-224.

De Vriend, H.J., Capobianco, M., Chesher, T., De Swart, H.E., Latteux, B., Stive, M.J.F., 1993b. Approaches to long term modeling of coastal morphology: a review. Coastal Engineering 21 (1-3), 225-269.

De Vriend, H.J., (1996), Mathematical modelling of meso-tidal barrier island coasts. Part I: Emperical and semi-emperical models, in Advances in coastal and ocean engineering, edited by Liu, P.L.F., World Scientific, Singapore, 115-149.

DiLorenzo, J.L., 1988. The overtide and filtering response of small inlet/bay systems. In: Aubrey, D.G. and Weishar, L., (Ed.), Hydrodynamics and Sediment Dynamics of Tidal Inlets, Springer, New York, USA, pp. 24-53.

Di Silvio, G., 1989. Modelling the morphological evolution of tidal lagoons and their equilibrium configurations. XXII Congress of the IAHR, Ottawa, Canada, 21-25 August 1989.

Dissanayake, D.M.P.K. and Roelvink, J.A. (2007), Process-based approach on tidal inlet evolution – Part 1, Proc. River, Coastal and Estuarine Morphodynamics: RCEM 2007, Dohmen-Janssen and Hulscher (eds), pp 3-9.

Dissanayake, M. van der Wegen and J.A. Roelvink (2009), *Modelled channel patterns in a schematized tidal inlet*, Coastal Engineering, doi: 10.1016/j.coastaleng.2009.08.008.

Dronkers, J. (1986), Tidal asymmetry and estuarine morphology, Neth. Journal of Sea Research 20, 117-131.

Dronkers, J. (1998), Morphodynamics of the Dutch Delta , In : Physics of Estuaries and Coastal Seas, edited by Dronkers, J. and Scheffers, M.B.A.M., Balkema, Rotterdam, pp297-304.

Dronkers, J. (2005), Dynamics of coastal systems. Advanced Series on Ocean Engineering, Vol 25, 520pp, World Scientific, Singapore.

Engelund, F. and Hansen, E., (1967), A monograph on sediment transport in alluvial streams, Teknisk Forlag, Copenhagen.

Escoffier, F.F., 1940. The stability of tidal inlets. Shore Beach 8 (4), pp. 114-115.

Eysink, W.D. (1990), Morphologic response of tidal basins to changes, in ASCE Proceedings of Coastal engineering Conference, Vol 2. 1948-1961, Delft, the Netherlands.

Feola, A., E.Belluco, A.D'Alpaos, S. Lanzoni, M, Marani., and A. Rinaldo, (2005) A geomorphic study of lagoonal landforms, Water Resources Research, Vol. 41, W06019, doi:10.1029/2004WR003811.

Fleming, K., P. Johnston, D. Zwartz, Y. Yokoyama, K. Lambeck and J. Chappell, (1998), Refining the eustatic sea-level curve since the Last Glacial Maximum using far- and intermediate-field sites, Earth and Planetary Science Letters 163 (1-4), 327-342.

Friedrichs, C.T. and D.G. Aubrey (1988), Non-linear tidal distortion in shallow well mixed estuaries: a synthesis, Estuarine, Coastal and Shelf Science, 27, 521-545.

Friedrichs, C.T., D.G. Aubrey and P.E., Speer (1990), Impacts of relative sea-level rise on evolution of shallow estuaries, In: Coastal and Estuarine Studies, edited by R.T. Cheng, vol 38, Residual currents and long-term transport, Springer-Verlag, New York, US, pp105-120.

Friedrichs, C.T. and D.G. Aubrey (1994), Tidal propagation in strongly convergent channels, Journal of Geophysical Research, 99(C2), 3321-3336.

Friedrichs, C.T., 1995. Stability shear stress and equilibrium cross-sectional geometry of sheltered tidal channels. Journal of Coastal Research, 11(4), pp 1062-1074.

Friedrichs, C.T. and D.G. Aubrey (1996), Uniform bottom shear stress and equilibrium hypsometry of intertidal flats, in: Mixing in estuaries and Coastal seas, edited by C. Pattiaratchi, Amer. Geoph. Union, Washington D.C., US.

Gallappatti, R.,and Vreugdenhil, C.B., 1985. A depth integrated model for suspended sediment transport. Journal of Hydraulic Research 23, 359-377.

Ganju, N.K., D.H. Schoellhamer, M.C. Murrell, J.W. Gartner and S.A. Wright. (2006), Constancy of the Relation Between Floc Size and Density in San Francisco Bay. In: Estuarine and Coastal Fine Sediment Dynamics, Volume 8: INTERCOH 2003 (Proceedings in Marine Science); p.75-91. Elsevier, New York, NY. Pp. 75-92. (ERL,GB 1209).

Ganju, N.K. and Schoellhamer, D.H., (2005), Lateral variability of the estuarine turbidity maximum in a tidal strait. In: Kusuda, T., Yamanishi, H., Spearman, J., and Gailani, J.Z., Eds., Sediment and Ecohydraulics: INTERCOH 2005, Elsevier, Amsterdam, the Netherlands.

Ganju, N.K., Schoellhamer, D.H., 2006. Annual sediment flux estimates in a tidal strait using surrogate measurements. Estuarine, Coastal, and Shelf Science 69, 165–178.

Ganju, N.K., Schoellhamer, D.H. 2007, Calibration of an estuarine sediment transport model to sediment fluxes as an intermediate step for simulation of geomorphic evolution, Continental Shelf Research, doi:10.1016/j.csr.2007.09.005.

Ganju, N.K., Schoellhamer, D.H., 2009. Calibration of an estuarine sediment transport model to sediment fluxes as an intermediate step for simulation of geomorphic evolution. Continental Shelf Research, 29. 148-158, doi:10.1016/j.csr.2007.09.005

Ganju, N.K., Schoellhamer, D.H., 2008. Lateral displacement of the estuarine turbidity maximum in tidal strait. In: Kusuda, T., Yamanishi, H., Spearman, J., Gailani, J.Z. (Eds.), Sediment and Ecohydraulics INTERCOH 2005, Proc. in Marine Science, Elsevier, Amsterdam, the Netherlands.

Ganju, N.K., Knowles, N., and Schoellhamer, D.H., in 2008, Temporal downscaling of decadal sediment load estimates to a daily interval for use in hindcast simulations. Journal of Hydrology, v. 349, p. 512-523.

Ganju, N. K., D. H. Schoellhamer, and B. E. Jaffe (2009), Hindcasting of decadal-timescale estuarine bathymetric change with a tidal-timescale model, J. Geophys. Res., 114, F04019, doi:10.1029/2008JF001191.

Gelfenbaum, G. , Roelvink, J.A., Meijs, M. and Ruggiero, P. 2003, process-based morphological modeling of Grays Harbor inlet at decadal time scales in: Proceedings of the fifth International Symposium on Coastal Engineering and Science of Coastal Sediment Processes, Coastal Sediments '03, ASCE, Clearwater Beach, Florida, USA, (available at CD-Rom published by World Scientific Publishing) ISBN 981-238-422-7

Gerritsen, F., De Jong, H., Langerak, A., 1990. Cross-sectional stability of estuary channels in the Netherlands. In: ASCE Proceedings of Coastal Engineering Conference, Delft, the Netherlands, Vol.2, pp. 2923-2935.

Gerritsen, F., Dunsbergen, D.W., Israel, C.G., 2003. A relational stability approach for tidal inlets, including analysis of the effect of wave action. Journal of Coastal research, 19 (4), pp. 1066-1081.

Gilbert, G.K., 1917. Hydraulic-mining debris in the Sierra Nevada: US Geological Survey Professional Paper 105, 148p.

Groen, P., (1967), On the residual transport of suspended matter by an alternating tidal current, Netherlands Journal of Sea Research, 3(4), 564-574.

Hampton, M.A., Snyder, N.P., Chin, J.L., Allison, D.W. and Rubin, D.M. 2003 Bed-Sediment Grain-Size and Morphologic Data from Suisun, Grizzly, and Honker Bays, CA, 1998-2002, USGS Open-File Report 03-250

Hayes, M.O., (1975), Morphology of sand accumulation in estuaries: an introduction to the symposium, in Estuarine research II, edited by L.E. Cronin, Academic Press, New York, 3-22.

Hayes, T.P., Kinney, J.J. and Wheeler, N.J.M (1984), California Surface wind climatology. California Air Resources Board, Aerometric Data Division, 73 pp

Hibma, A., H.M. Schuttelaars and Z.B. Wang (2003a), Comparison of longitudinal equilibrium profiles in idealized and process-based models, Ocean Dynamics, Vol. 53 (3) pp. 252-269.

Hibma, A., H.J. de Vriend and M.J.F. Stive (2003b), Numerical modeling of shoal pattern formation in well-mixed elongated estuaries. Estuarine, Coastal and Shelf Science. Vol. 57, 5-6, pp. 981-991.

Hibma, A., H.M. Schuttelaars and H.J. de Vriend (2003c), Initial formation and long-term evolution of channel–shoal patterns, Continental Shelf Research, Vol: 24, Issue 15, doi:10.1016/j.csr.2004.05.003

Hibma, M.J.F. Stive, Z.B. Wang, (2004a). Estuarine morphodynamics, Coastal Engineering, Volume 51, Issues 8-9, Coastal Morphodynamic Modeling, Pages 765-778, DOI: 10.1016/j.coastaleng.2004.07.008.

Hibma, A., Wang, Z.B., Stive, M.J.F. and De Vriend, H.J., (2004b). Modelling impact of dredging and dumping in ebb-flood channel systems. Proc. Of Physics of Estuaries and Coastal seas, Yucatán, Mexico

Hibma, A., 2004. Morphodynamic modeling of estuarine channel-shoal systems. Ph.D. thesis, Technical University Delft, The Netherlands

Higgins, S.A., Jaffe, B.E., and Fuller, C.C., 2007, Reconstructing sediment age profiles from historical bathymetry changes in San Pablo Bay, California, Estuarine Coastal and Shelf Science, doi:10.1016/j.ecss.2006.12.018

Hoitink, A.J.F., Hoekstra, P., and van Maren, D.S., 2003. Flow asymmetry associated with astronomical tides: Implications for the residual transport of sediment, J. Geophysical Research, 108(C10), 3315.

Horrevoets, A.C., Savenije, H.H.G., Schuurman, J.N. and Graas, S., 2004. The influence of river discharge on tidal damping in alluvial estuaries. Journal of Hydrology 294, 213-228

Horton, R.E. (1945), Erosional development of streams and their drainage basins: Hydrophysical approach to quantitative morphology, Geol. Soc. Am. Bull., 56, 275-370.

Hughes, S.A., 2002. Equilibrium cross-sectional are at tidal inlets. Journal of Coastal Research, 18 (1), pp. 160-174.

Hume, T.M., Herdendorf, C.E., 1993. On the use of empirical stability relationships for characterising estuaries. Journal of Coastal Research, 9(2), pp. 413-422.

Ikeda, S., G. Parker, and K. Sawai, (1981), Bend theory of river meanders. part 1. Linear development, Journal of Fluid Mechanics, Vol. 112, 363-377.

Ikeda, S., (1982), Lateral bed load transport on side slopes, Journal Hydraulics Division, ASCE, Vol. 108, no.11, 1369-1373.

Ikeda, S., and T. Aseada, (1983), Sediment suspension with rippled bed, Journal Hydraulics Division, ASCE, Vol. 109, no.3, 409-423.

Jaffe, B.E., Smith, R.E., Zink Toressan, L., 1998. Sedimentation and bathymetric change in San Pablo Bay 1856-1983. USGS Open File Report, 98-0795.

Jaffe, B.E., Smith R.E. and Foxgrover A.C., 2007. Anthropogenic influence on sedimentation and intertidal mudflat change in San Pablo Bay, California: 1856-1983. Estuarine, Coastal and Shelf Science 73 175-187.

Jarrett, J.T. (1976), Tidal prism-inlet relationships, Gen. Invest. Tidal inlets Rep. 3, 32 pp, US Army Coastal Engineering and Research Centre. Fort Belvoir, Va.

Jay, D.A., (1991). Green's law revisited: Tidal long wave propagation in channels with strong topography, Journal of Geophysical Research, Vol. 96, no C11, 20,585-20,598

Jeuken, M.C.J.L., 2000. On the morphologic behaviour of tidal channels in the Westerschelde estuary. Ph.D. thesis, Utrecht University, the Netherlands

Jones, C. (2008), Aquatic Transfer Facility Sediment Transport Analysis, In: Technical Studies for the Aquatic Transfer Facility: Hamilton Wetlands Restoration Project, edited by Cacchione, D.A., and Mull, P.A., U.S. Army Corps of Engineers, San Francisco District, Hamilton Wetland Restoration Project Aquatic Transfer Facility Draft Supplemental EIS/EIR, Appendix A, Chapter 4, pp. 305-340.

http://www.hamiltonwetlands.org/hw_media/docs/Hamilton ATF SEIS Appendices.pdf

Keulegan, G.H., 1967. Third progress report on tidal flow in entrances: water level fluctuations of basins in communication with seas. Report no:1146, National Bureau of standards, Washington DC, USA,

Kimmerer, W.J. 2004. Open Water processes of the San Francisco Estuary: from physical forcing to biological responses. San Francisco Estuary and Watershed Science [online serial]. Vol.2, Issue 1, art 1. http://repositories.cdlib.org/jmie/sfews/vol2/iss1/art1

Kineke, G.C., and Sternberg, R.W., 1989, The effect of particle settling velocity on computed suspended sediment concentration profiles: Marine Geology, v. 90, p. 159-174.

Krank, K., and Milligan, T.G., 1992, Characteristics of suspended particles at an 11-hour anchor station in San Francisco Bay, California: Journal of Geophysical Research, v. 97, no. C7, p. 11373-11382.

Kraus, N.C., 1998. Inlet cross-sectional area calculated by process-based model. In: Proceedings of the 26th Conference on Coastal Engineering, vol.3, ASCE, pp. 3265-3278.

Krishnamurthy, M., 1977. Tidal prism of equilibrium inlets. Journal of the Waterways, Harbors and Coastal Engineering Division, ASCE, 103(WW4), pp. 423-432

Krone, R.B., (1962) Flume studies of transport of sediment in estuarine shoaling processes, Final Report, Hydraulics Engineering Research Laboratory, Un. of California, Berkeley, CA.

Krone, R.B., 1979, Sedimentation in the San Francisco Bay system. In: Conomos, T.J. (Ed.), San Francisco Bay: The Urbanized Estuary. Pacific Division of the American Association for the Advancement of Science, San Francisco, California, pp. 85–96.

Kuijper, C., R. Steijn, D. Roelvink, T. van der Kaaij,and P. Olijslagers, (2004), Morphological modelling of the Western Scheldt. Validation of Delft3D, WL|Delft Hydraulics / Alkyon, Report Z3648.

Lane, A. (2004), Bathymetric evolution of the Mersey estuary, UK, 1906-1997: causes and effects, Estuarine, Coastal and Shelf Science, 59, 249-263.

Lanzoni, S. and Seminara, G., 1998. On tide propagation in convergent estuaries. Journal of Geophysical Research, 103(C13), 30793-30812.

Lanzoni, S. and G. Seminara (2002), Long-term evolution and morphodynamic equilibrium of tidal channels, Journal of Geophysical Research, 107 (C1), 3001, doi:10.1029/2000JC000468.

Langbein, W.B. (1963), The hydraulic geometry of a shallow estuary, Bulletin of Internat. Association of Scientific Hydrology, 8, 84-94.

Latteux, B. (1995), Techniques for long-term morphological simulation under tidal action, Marine geology 126, 129-141.

LeConte, L.J., 1905. Discussion on river and harbour outlets. "Notes on the improvement of river and harbour outlets in the United States", Paper nr 1009, by D.A. Watts, Trans., ASCE 55, pp. 306-308.

Leendertse, J.J. (1987), A three-dimensional alternating direction implicit model with iterative fourth order dissipative non-linear advection terms, WD-3333-NETH, Rijkswaterstaat, The Hague, The Netherlands.

Leopold, L.B. and W.B. Langbein (1962), The concept of entropy in landscape evolution, Geological Survey Professional Paper, 50-A, A1-A20.

Lesser, G.R., Roelvink, J.A., Van Kester, J.A.T.M., Stelling, G.S., 2004. Development and validation of a three-dimensional morphological model. Coastal Engineering 51, pp. 883-915.

Lesser, G.R., 2009. An approach to medium-term coastal morphological modelling. PhD thesis, UNESCO-IHE & Delft Technical University, Delft. CRC Press/Balkema. ISBN 978-0-415-55668-2.

Locke, J.L., 1971. Sedimentation and foraminiferal aspects of the recent sediments of San Pablo Bay. MSC thesis, Fac. of the Dep. of Geology, San Jose State College.

Marani, M., Lanzoni, S., Zandolin, D., 2002. Tidal meaders. Water Resources Research, Vol. 38, No 11, 1225, doi:10.1029/2001WR000404

Marciano, R., Z.B. Wang, A. Hibma, H.J. de Vriend and A. Defina, (2005), Modelling of channel patterns in short tidal basins, Journal of Geophysical Reserach, Vol.100, F01001, doi:10.1029/2003JF000092.

Mason,C. 1973. Regime equations and tidal inlets. Journal of the Waterways, Harbors and Coastal Engineering Division, ASCE, 99(WW3), pp. 393-397.

McDonald, E.T., Cheng, R.T., 1997. A numerical model of sediment transport applied to San Francisco Bay, California. Journal of Marine Environmental Engineering 4, pp. 1–41.

McKee, L.J., Ganju, N.K., Schoellhamer, D.H., 2006. Estimates of suspended sediment entering San Francisco Bay from the Sacramento and San Joaquin Delta, San Francisco Bay, California. Journal of Hydrology 323, 335–352.

Mehta, A.J., and Özsoy, E., 1978. Flow dynamics and near shore transport. In: Stability of tidal inlets, Developments in Geotechnical Engineering 23, edited by Bruun, P., Elsevier, Amsterdam, 83-161.

Monismith, S.G., Kimmerer, W., Burau, J.R., Stacey, M.T., 2002. Structure and flow-induced variability of the subtidal salinity field in Northern San Francisco Bay. Journal of Physical Oceanography, Vol.32, pp 3003-3019.

Muller, G., and U. Forstner, 1968, General relationship between suspended sediment concentration and water discharge in the Alpenrhein and some other rivers, Nature, 217, 244-245.

Murphy, A.H., and E.S. Epstein, Skill scores and correlation coefficients in model 890 verification, *Monthly Weather Review, 117*, 572–581, 1989.

Murray, A.B., M.A.F. Knaappen, M. Tal, and M.L. Kirwan, (2007), Biomorphodynamics in river, coastal and estuarine settings, in Proceedings of 5thIAHR Conference on River, Coastal and Estuarine Morphodynamics (RCEM 2007), Enschede, the Netherlands, 233-246.

Nederbragt, G. and G.J. Liek, (2004), *In Dutch,* Description of sand balance Westerschelde estuary and river mouth [Beschrijving zandbalans Westerschelde en monding]. Rapport RIKZ, 2004.020. RIKZ: Den Haag, The Netherlands. 70 pp. Available at http://www.vliz.be/imis/imis.php?module=ref&refid=64651

O'Brien, M.P., 1931. Estuary and tidal prisms related to entrance areas. Civil Engineering, Vol 1, 8, pp. 738-739.

O'Brien, M.P., (1969), Equilibrium flow areas of inlets on sandy coasts, Journal of the Waterways and Harbours Division, ASCE. 95(WW1), pp. 43-51.

Parker, B.B. (Ed.), (1991), Tidal hydrodynamics, John Wiley & Sons, New York.

Partheniades, E., 1965. Erosion and Deposition of Cohesive Soils. Journal of the Hydraulic Division, ASCE, Vol 91, No. HY1

Pillsbury, G., (1956), Tidal Hydraulics, Corps of Engineers, Vicksburg, USA

Pethick, J.S., (1994), Estuaries and wetlands: function and form, in Wetland Management, Thomas Telford, London, 75-87

Porterfield, G., 1980, Sediment transport of streams tributary to San Francisco, San Pablo, and Suisun Bays, California, 1909–1966: U.S. Geological Survey Water-Resources Investigations Report 80-64, 91 p.

Prandle, D., (2003), Relationship between tidal dynamics and bathymetry in strongly convergent estuaries, Journal of Physical Oceanography, 33, 2738-2750.

Prandle, D., (2004), Sediment trapping, turbidity maxima, and bathymetric stability in macrotidal estuaries, Journal of Geophysical Research, 109, C08001, oi:10.1029/2004JC002271.

Prandle, D., A. Lane, and A. J. Manning (2005), Estuaries are not so unique, Geophys. Res. Lett., 32, L23614, doi:10.1029/2005GL024797

Pritchard, D.W., (1952), Estuarine hydrography, in Advances in Geophysics, Vol. 1, edited by H. E. Landsberg, Academic Press, Inc., New York, USA, 243-245.

Pritchard, D. (2005), Suspended sediment transport along an idealized tidal embayment: settling lag, residual transport and the interpretation of tidal signals, Ocean Dynamics, 55, 124-136.

Rees, J.G., J. Ridgway, S. Ellis, R.W.O'B. Knox, R. Newsham, A. Parkes, (2000), Holocene sediment storage in the Humber Estuary, in Holocene land-ocean interaction and environmental change around the North Sea, edited by I. Shennan and J.E. Andrews, Geological Society, London, pp. 119-143.

Reynolds, O. (1887), On certain laws relating to the regime of rivers and estuaries and on the possibility of experiments on a small scale, Br. Assoc. Rep., pp. 555–562, London.

Reynolds, O. (1889), On model estuaries I, On the action of waves and currents, Report of British Association for the advancement of science, pp 327-343, Br. Assoc., London.

Reynolds, O. (1890), On model estuaries II, On the action of waves and currents, Report of British Association for the advancement of science, pp 512-534, Br. Assoc., London.

Reynolds, O. (1891), On model estuaries III, On the action of waves and currents, Report of British Association for the advancement of science, pp 386-404, Br. Assoc., London.

Rinaldo, A., S. Fagherazzi, S., S. Lanzoni, M. Marani, and W.E. Dietrich (1999a), Tidal networks 2. Watershed delineation and comparative network morphology, Water resources Research, Vol.35, no. 12, 3905-3917.

Rinaldo, A., S. Fagherazzi, S., S. Lanzoni M. Marani, and W.E. Dietrich (1999b), Tidal networks 3. Landscape-forming discharges and studies in empirical geomorphic relationships, Water resources Research, Vol.35, no. 12, 3919-3929.

Rinaldo, A., J.R. Banavar and A. Maritan, (2006), Trees, networks, and hydrology, Water Resources Research, 42, W06D07, doi: 10.1029/2005WR004108

Rodriguez-Iturbe, I., A. Rinaldo, R. Rigon, R.L. Bras, A. Marani, and E. Ijjasz-Vasquez, (1992), Energy dissipation, runoff production, and the three-dimensional structure of river basins, Water Resources Research, 28, No. 4, 1095-1103.

Rodriguez-Iturbe, I., and Rinaldo, A., 1997. Fractal River Basins, Chance and Self-Organisation. Cambridge University Press, UK.

Roelvink, J.A., T. van Kessel, S. Alfageme and R. Canizares, (2003), Modelling of barrier island response to storms, in: Proceedings of the fifth International Symposium on Coastal Engineering and Science of Coastal Sediment Processes, Coastal Sediments '03, ASCE, Clearwater Beach, Florida, USA, (available at CD-Rom published by World Scientific Publishing).

Roelvink, J.A. and Walstra D. (2004), Keeping it simple by using complex models. Advances in Hydro-Science and Engineering, Volume VI, pp1-11

Roelvink, J.A. (2006), Coastal morphodynamic evolution techniques, Journal of Coastal Engineering, Vol. 53, 177-187.

Rubin, D.M. and David S. McCulloch. 1979. The Movement and Equilibrium of Bedforms in Central San Francisco Bay. *In*: Conomos, T. J., editor. San Francisco Bay: The Urbanized Estuary. http://www.estuaryarchive.org/archive/conomos_1979

Ruhl, C.A., Schoellhamer, D.H., Stumpf, R.P., and Lindsay, C.L., 2001, Combined use of remote sensing and continuous monitoring to analyse the variability of suspended-sediment concentrations in San Francisco Bay, California: Estuarine, Coastal and Shelf Science, v. 53, p. 801-812.

Ruhl, C.A., and Schoellhamer, D.H. 2004. Spatial and temporal variability of suspended-sediment concentration in a shallow estuarine environment. San Francisco Estuary and Watershed Science [online serial]. Vol. 2, Issue 2 (May 2004), Article 1. http://repositories.cdlib.org/jmie/sfews/vol2/iss2/art1

Savenije, H. H.G., 1992a. Rapid assessment technique for salt intrusion in alluvial estuaries. PhD thesis, IHE report series 27, UNESCO-IHE, Delft.

Savenije, H.H.G., 1992b. Lagrangian solution of St Venants equations for an alluvial estuary. Journal of Hydraulic Engineering, ASCE, 118(8), 1153-1163.

Savenije, H.H.G, 1998. Analytical expression for tidal damping in alluvial estuaries. Journal of Hydraulic Engineering, ASCE, 124 (6), 615-618

Savenije, H.H.G., (2001), A simple analytical expression to describe tidal damping or amplification, Journal of Hydrology, 243, 205-215.

Savenije, H.H.G., 2003. The width of a bankfull channel: Lacey's formula explained. Journal of Hydrology 276, pp. 176-183.

Savenije, H.H.G. and E.J.M. Veling, (2005), Relation between tidal damping and wave celerity in estuaries, Journal of Geophysical Research, 110, C04007, doi:10.1029/2004JC002278.

Savenije, H. H.G. (2005), Salinity and tides in alluvial estuaries, pp 194, Elsevier. Amsterdam, the Netherlands

Savenije, H. H. G., M. Toffolon, J. Haas, and E. J. M. Veling (2008), Analytical description of tidal dynamics in convergent 21 estuaries, J. Geophys. Res., 113, doi:10.1029/2007JC004408

Schoellhamer, D.H., Burau, J.R., 1998. Summary of findings about circulation and the estuarine turbidity maximum in Suisun Bay, California. US Geological Survey Fact Sheet FS-047-98, 6 pp.

Schoellhamer, D.H., 2002, Comparison of basin-wide effect of dredging and natural estuarine processes on suspended-sediment concentration: Estuaries, v. 25, no. 3, p. 488-495.

Schoellhamer, D.H., Ganju, N.K., and Shellenbarger, G.G., 2008, Sediment Transport in San Pablo Bay, in Cacchione, D.A., and Mull, P.A., eds., Technical Studies for the Aquatic Transfer Facility: Hamilton Wetlands Restoration Project: U.S. Army Corps of Engineers, San Francisco District, Hamilton Wetland Restoration Project Aquatic Transfer Facility Draft Supplemental EIS/EIR, Appendix A, Chapter 2, p. 36-107.
http://www.hamiltonwetlands.org/hw_media/docs/Hamilton   ATF   SEIS Appendices.pdf

Schramkowski, G.P., H.M. Schuttelaars and H.E. De Swart (2002), The effect of geometry and bottom friction on local bed forms in a tidal embayment, Continental Shelf Research 22, 1821-1833

Schramkowski, G.P., H. M. Schuttelaars and H.E. De Swart (2004), Non-linear channel-shoal dynamics in long tidal embayments, Ocean Dynamics 54, 399-407

Schuttelaars, H.M. and H.E. De Swart (1996), An idealized long term morphodynamic model of a tidal embayment, European Journal of Mechanics, B/Fluids 15 (1), 55-80

Schuttelaars, H.M. and H.E. De Swart (1999), Initial formation of channels and shoals in a short tidal embayment, Journal of Fluid Mechanics 386, 15-42

Schuttelaars, H.M. and H.E. De Swart (2000), Multiple morphodynamic equilibria in tidal embayments, Journal of Geophysical Research 105, No C10, 24,105-24,118

Seabergh, W.C. 2003. Long-term coastal inlet channel area stability, Proc. Coastal Sediments '03. 2003. CD-ROM Published by World Scientific Publishing Corp., Corpus Christi, Texas, USA.

Singh, V.P., C.T Yang and Z-Q Deng, (2003), Downstream hydraulic geometry relations: 1.Theoretical development, Water Resources Research, 39(12),1337, doi:10.1029/2003WR002484.

Seminara, G., and M. Tubino (2001), Sand bars in tidal channels. Part 1. Free bars. J. Fluid Mech. 440, 49-74.

Seminara, G., Blondeaux, P., 2001. Perspectives in morphodynamics. In: River, coastal and estuarine morphodynamics, Springer Verlag, Berlin.

Sha, L.S. and J.H. Van den Berg (1993), Variation in ebb-tidal delta geometry along the coast of the Netherlands and the German Bight,. Journal of Coastal Research, Vol 9, 730-746.

Smith, R. A., 1980: Golden Gate tidal measurements: 1854–1978. J.Waterway, Port, Coastal Ocean Div., Proc. Amer. Soc. Civ. Eng.,106, 407–410.

Solari, L., Seminara, G., Lanzoni, S., Marani, M. and Rinaldo, A. 2001 Sand bars in tidal channels. Part 2. Tidal meanders. Submitted for publication J. Fluid Mech.

Speer, P.E. and D.G. Aubrey (1985), A study of non-linear propagation in shallow inlet/estuarine systems. Part II;theory, Estuarine, Coastal and Shelf Science, 21, 207-224

Stelling, G.A. (1984), On the construction of computational methods for shallow water flow problems, Rijkswaterstaat Communications, No 35, Rijkswaterstaat, The Hague, the Netherlands

Stelling, G.A. and J.J. Leendertse (1991), Approximation of convective processes By cyclic ACI methods, in: Proceedings 2nd ASCE Conference on estuarine and coastal modelling, Tampa, United States of America

Stive, M.J.F., Capobianco, M., Wang, Z.B., Ruol, P., Buijsman, M.C., 1998. Morphodynamics of a tidal lagoon and the adjacent coast. In: Dronkers, J., Scheffers, M. (eds), Physics of estuaries and coastal seas. Balkema, Rotterdam.

Stive, M.J.F. and Wang, Z.B., 2003. Morphodynamic modeling of tidal basins and coastal inlets. In: Lakhan, V.C. ed., Advances in coastal modeling, Elsevier, 367-392

Strahler, A.N., 1952. Hypsometric (area-altitude) analysis of erosional topography. Geollogical Society Am. Bull. 63: 1117-1142

Struiksma, N.K., K.W. Olesen, C. Flokstra, and H.J. De Vriend (1985), Bed deformation in curved alluvial channels, Journal of Hydraulic Research, Vol. 23, 57-79.

Sutherland, J., Peet, A.H., and Soulsby R.L. (2004), Evaluating the performance of morphological models, Coastal Engineering, doi:10.1016/j.coastaleng.2004.07.015

Sternberg, D.W., Cacchione, D.A., Drake, D.E., Kranck, K. 1986. Suspended sediment transport in an estuarine tidal channel within San Francisco Bay, California, Marine Geology, 71, 00 237-258.

Tambroni, N., M. Bolla Pittaluga and G. Seminara (2005), Laboratory observations of the morphodynamic evolution of tidal channels and tidal inlets, Journal of Geophysical Research, Vol. 110, F04009, doi: 10.1029/2004JF000243

Tambroni, N. and G Seminara, 2010. On the theoretical basis of O'Brien-Jarrett-Marchi law for tidal inlets and tidal channels. in: Proc. 6thIAHR (RCEM 2009, edited by Vionett et al., pp 329-335, Taylor and Francis Group, London. ISBN 978-0-415-55246-8.

Teeter, A.M. 1987 Alcatraz disposal site investigation, San Francisco Bay; Alcatraz disposal site erodibility, Report 3. Misc Paper HL-86-1, US Army Eng. Waterway Exp. Station, Vicksburg, MS.

Toffolon, M., G. Vignoli, and M. Tubino (2006), Relevant parameters and finite amplitude effects in estuarine hydrodynamics, J. Geophys. Res., 111, C10014, doi:10.1029/2005JC003104.

Toffolon, M., and A. Crosato (2007),‡Developing Macroscale Indicators for Estuarine Morphology:The Case of the Scheldt Estuary, Journal of Coastal Research, vol 23(1), 195-212.

Townend, I.H., (1999), Long-term changes in estuary morphology using the entropy method, in IAHR Symposium on River, Coastal and Estuarine Morphodynamics, University of Genova, Department of Environmental Engineering, II, 339-348.

Townend, I.H., and R.W. Dun, (2000), A diagnostic tool to study long-term changes in estuary morphology, in Coastal and Estuarine Environments, sedimentology, geomorphology and geoarcheology, edited by K. Pye and J.R.L. Allen,Geological Society, London, 75-86.

Uncles, R.G., and Peterson, D.H., 1996. The long-term salinity field in San Francisco Bay. Continental Shelf Research, Vol. 16, no. 15, pp. 2005-2039.

Van den Berg, J.H., Jeuken, M.C.J.L., Van der Spek, A.J.F., 1996. Hydraulic processes affecting the morphology and evolution of the Westerschelde estuary. In: Nordstrom, K.F., Roman, C.T., eds., Estuarine shores: Evolution, Environments and Human Alterations. London, Johm Wiley, 157-184

Van de Kreeke, J. and K. Robaczewska, (1993), Tide induced residual transport of coarse sediment; application to the Ems estuary, Netherlands Journal of Sea Research, 31 (3), Netherlands Institute for Sea Research, 209-220.

Van de Kreeke, J., 1998. Adaptation of the Frisian Inlet to a reduction in basin area with special refrence to the cross-sectional area of the inlet channel, In : Physics of Estuaries and Coastal Seas, edited by Dronkers, J. and Scheffers, M.B.A.M., Balkema, Rotterdam, the Netherlands, pp. 355-362.

Van de Kreeke, J., 2004. Equilibrium and cross-sectional stability of tidal inlets: application to the Frisian inlet before and after basin reduction, Coastal Engineering, 51, pp. 337-350.

Van de Kreeke J. 2006. An aggregate model for the adaption of the morphology and sand bypassing after basin reduction of the Frisian Inlet. Coastal Engineering 53:255–263

Van der Spek, A.J.F., (1997), Tidal asymmetry and long-term evolution of holocene tidal basins in The Netherlands: simulation of palaeo-tides in the Schelde estuary, Marine Geology, 141, 71-90.

Van der Wegen, M., D.Q. Thanh, and J.A. Roelvink, (2006), Bank erosion and morphodynamic evolution in alluvial estuaries using a process-based 2D model, in Proceedings ICHE conference 2006, Philadelphia, USA (available at http://hdl.handle.net/1860/1475)

Van der Wegen, M., Z.B. Wang, H.H.G. Savenije and J.A. Roelvink, (2007), Estuarine evolution using a numerical, process-based approach, in Proceedings of 5thIAHR Conference on River, Coastal and Estuarine Morphodynamics (RCEM 2007), Enschede, the Netherlands, 95-101.

Van der Wegen, M., and J. A. Roelvink (2008), Long-term morphodynamic evolution of a tidal embayment using a two-dimensional, process-based model, J. Geophys. Res., 113, C03016, doi:10.1029/2006JC003983.

Van der Wegen, M., Z. B. Wang, H. H. G. Savenije, and J. A. Roelvink (2008), Long-term morphodynamic evolution and energy dissipation in a coastal plain, tidal embayment, J. Geophys. Res., 113, F03001, doi:10.1029/2007JF000898

Van der Wegen, M., Z.B. Wang, I.H. Townend, H.H.G. Savenije and J.A. Roelvink, (2009), Long-term, morphodynamic modeling of equilibrium in an alluvial tidal basin using a process-based approach, in Proceedings of 6th IAHR Conference on River, Coastal and Estuarine Morphodynamics (RCEM 2009), Santa Fé, Argentina, pp. 229-321.

Van Dongeren, A.D. and H.J. De Vriend (1994), A model of morphological behaviour of tidal basins, Coastal Engineering, 22, 287-310

Van Geen, and S.N. Luoma. 1999. Impact of human activities on sediments of San Francisco Bay, California: an overview, Marine Chemistry 64 1999 1–6.

Van Leeuwen, S.M. and De Swart, H.E., Effect of advective and diffusive sediment transport on the formation of local and global patterns in tidal embayments, Ocean Dynamics 54, 441-451, 2004

Van Rijn, L.C. (1993), Principles of sediment transport in rivers, estuaries and coastal seas, pp 535, AQUA Publications, the Netherlands

Van Rijn, L.C., Walstra, D.J.R., Grasmeijer, B., Sutherland, J., Pan, S., Sierra, J.P., 2003. The predictability of cross-shore bed evolution of sandy beaches at the time scale of storms and seasons using process-based profile models. Coastal Engineering 47, 295– 327.

Van Veen, J. (1936), Onderzoekingen in de Hoofden (in Dutch), Rijksuitgeverijdienst van de Ned. Staatscourant, The Hague, Netherlands.

Van Veen, J. (1950), Eb en vloed scharen in de Nederlandsche getij wateren, J. R. Dutch Geogr. Soc., 67, 303 –325. (Engl. transl., Delft Univ. Press, Delft, Netherlands, 2002.)

Van Veen, J., 2002. Ebb and Flood channel system in the Netherlands tidal waters, English translation of original Dutch text, Delft University Press, Delft, the Netherlands (original Dutch text: Van Veen, J., 1950. Eb en vloed scharen in de Nederlandsche getij wateren. Journal of the Royal Dutch Geographical Society (KNAG), 67, 303-325)

Walton, T.L. Jr. and Adams, W.D., 1976. Capacity of inlet outer bars to store sand. Proceedings Coastal Engineering Conference Honolulu, ASCE, 1919-1937

Walton, T.L., 2004. Escoffier curves and inlet stability. Journ. of Waterway, Port, Coastal and Ocean Engineering, Vol. 130, iss. 1, pp. 54-57, doi 10.1061/(ASCE)0733-950X(2004)130:1(54).

Wang, Z.B., Louters, T. and De Vriend, H.J. (1991), A morphodynamic model for a tidal inlet, in: Computer Modelling in Ocean Engineering '91, edited by Arcilla, A.S. et al, Balkema, Rotterdam, the Netherlands, pp 235-245.

Wang, Z. B., C. Jeuken, and H. J. De Vriend (1999), Tidal asymmetry and residual sediment transport in estuaries. A literature study and applications to the Western Scheldt, Rep. Z2749, WL|Delft Hydraul., Delft, Netherlands.

Winterwerp, J.C., Z.B. Wang, M.J.F. Stive, A. Arends, C. Jeuken, C. Kuijper and P.M.C. Thoolen, (2001), A new morphological schematization of the Western Scheldt Estuary, The Netherlands, in Proceedings of 2nd IAHR Symposium on River, Coastal and Estuarine Morphodynamics RCEM (2001), Obihiro, Japan, pp. 525-533.

Winterwerp, J.C. and Van Kesteren, W.G.M. (2004), Introduction to the physics of cohesive sediment in the marine environment. Developments in sedimentology 56, Elsevier, Amsterdam, 466 p.

Woodroffe, C.D., M.E. Mulrennan and J. Chappell, (1993), Estuarine infill and coastal progradation, southern van Diemen Gulf, northern Australia, Sedimentary Geology, 83, 257-275.

Wright, S.A., Schoellhamer, D.H. 2004. Trends in the sediment yield of the Sacramento River, California, 1957-2001. San Francisco Estuary and Watershed Science [online serial]. Vol.2, Issue 2, art 2.
http://repositories.cdlib.org/jmie/sfews/vol2/iss2/art2

Yalin, M.S. and A.M. Da Silva (1992), Horizontal turbulence and alternate bars, Journal of Hydroscience and Hydraulics Engineering, Vol. 9, no.2, 47-58.

Yang, Z. and Hamrick, J.M. (2003), Variational inverse parameter estimation in a cohesive sediment transport mode: An adjoint approach, Journ. of Geoph. Res., Vol. 108, 52, 3055, doi:10.1029/2002JC001423.

List of symbols

Symbol	Unit	Meaning
C	$[\mathrm{m}^{1/2}/\mathrm{s}]$	friction parameter defined by $\dfrac{\sqrt[6]{h}}{n}$
c_f	[-]	friction coefficient
D_{50}	[m]	median grain size
g	$[\mathrm{m}^2/\mathrm{s}]$	gravitational acceleration
h	[m]	water depth
n	$[\mathrm{sm}^{-1/3}]$	Manning's coefficient
r	[-]	correlation coefficient
S	$[\mathrm{m}^3/\mathrm{ms}]$	magnitude of the sediment transport per meter width
S_b	$[\mathrm{m}^3/\mathrm{ms}]$	magnitude of bed load transport per meter width
S_s	$[\mathrm{m}^3/\mathrm{ms}]$	magnitude of suspended transport per meter width
S_x	$[\mathrm{m}^3/\mathrm{ms}]$	sediment transport in x-direction per meter width
S_y	$[\mathrm{m}^3/\mathrm{ms}]$	sediment transport in y-direction per meter width
U	[m/s]	magnitude of flow velocity
$\bar{u}$	[m/s]	depth averaged velocity in x direction,
$\bar{v}$	[m/s]	depth averaged velocity in y direction,
z_b	[m]	bed level
α_{bs}	[-]	longitudinal bed slope coefficient (default 1)
α_{bn}	[-]	lateral bed slope coefficient (default 1.5)
Δ	[-]	relative density $(\rho_s - \rho_w)/\rho_w$
ε	[-]	bed porosity (default 0.4)
ζ	[m]	water level with respect to datum
ν_e	$[\mathrm{m}^2/\mathrm{s}]$	eddy viscosity
ρ_s	$[\mathrm{kg}/\mathrm{m}^3]$	sand density
ρ_w	$[\mathrm{kg}/\mathrm{m}^3]$	water density
σ	[-]	standard deviation
ϕ	$[^0]$	internal angle of friction (default 30^0)

List of symbols

Symbol	Unit	Meaning
a	$[m^2/s]$	friction parameter defined by $a = \frac{U}{c^2h}$
c	$[-]$	friction coefficient
D_{50}	$[m]$	median grain size
g	$[m/s^2]$	gravitational acceleration
h	$[m]$	water depth
n	$[s m^{-1/3}]$	Manning's coefficient
r	$[-]$	correlation coefficient
S	$[m^2/s]$	magnitude of the sediment transport per meter width
S_b	$[m^2/s]$	magnitude of bed load transport per meter width
S_s	$[m^2/s]$	magnitude of suspended transport per meter width
S_x	$[m^2/s]$	sediment transport in x-direction per meter width
S_y	$[m^2/s]$	sediment transport in y-direction per meter width
U	$[m/s]$	magnitude of flow velocity
u	$[m/s]$	depth-averaged velocity in x-direction
v	$[m/s]$	depth-averaged velocity in y-direction
z_b	$[m]$	bed level
α_s	$[-]$	longitudinal bed slope coefficient (default 1)
α_n	$[-]$	lateral bed slope coefficient (default 1.5)
Δ	$[-]$	relative density $\Delta = (\rho_s - \rho_w)/\rho_w$
ε	$[-]$	bed porosity (default 0.4)
ζ	$[m]$	water level with respect to depth
ν	$[m^2/s]$	eddy viscosity
ρ_s	$[kg/m^3]$	sand density
ρ_w	$[kg/m^3]$	water density
σ	$[-]$	standard deviation
φ	$[°]$	internal angle of friction (default 30)

Exposure

Peer-reviewed publications

Dissanayake, M. van der Wegen and J.A. Roelvink (2009), *Modelled channel patterns in a schematized tidal inlet*, Coastal Engineering, doi: 10.1016/j.coastaleng.2009.08.008.

Van der Wegen, M., Z. B. Wang, H. H. G. Savenije, and J. A. Roelvink, (2008), *Long-term morphodynamic evolution and energy dissipation in a coastal plain, tidal embayment*, J. Geophys. Res., 113, F03001, doi:10.1029/2007JF000898.

Van der Wegen, M., and J. A. Roelvink (2008), *Long-term morphodynamic evolution of a tidal embayment using a two-dimensional, process-based model*, J. Geophys. Res., 113, C03016, doi:10.1029/2006JC003983.

Publications submitted or in preparation

Van der Wegen, M., H.H.G. Savenije and J. A. Roelvink (in prep.), *Development of equilibrium estuarine geometry.*

Van der Wegen, M., B.E. Jaffe and J. A. Roelvink, (submitted), *Process-based, morphodynamic hindcast of decadal deposition patterns in San Pablo Bay, California, 1856-1887 .*

Van der Wegen and J. A. Roelvink, (in prep.) *Reproduction of the Western Scheldt bathymetry by means of a process-based, morphodynamic model*

Van der Wegen, M., B.E. Jaffe and J. A. Roelvink, (submitted to Ocean Dynamics), *Generation of bed composition for hindcasting morphodynamic evolution in San Pablo Bay, California*

Van der Wegen, M.,A. Dastgheib and J. A. Roelvink, (submitted to Coastal Engineering), *Morphodynamic modelling of tidal channel evolution in comparison to empirical PA relationship*

Conference attendance and proceedings

2009   Mick van der Wegen, Zheng Bing Wang, Ian Townend, Huub Savenije and Dano Roelvink, *Long-term, morphodynamic modeling of equilibrium in an alluvial tidal basin using a process-based approach*, proceedings RCEM conference, Santa Fé, Argentina (oral presentation).

2009   Komla W. Ofori, Mick van der Wegen, Dano Roelvink and John de Ronde, *Investigating the trends of import and export of sediment in the Western Scheldt estuary*, the Netherlands - by a process-based model, proceedings RCEM conference, Santa Fé, Argentina (poster).

2009   Mick van der Wegen, Bruce Jaffe, Ali Dastgheib, Dano Roelvink, *Generation of initial bed composition for morphodynamic hindcasting of hydraulic mining deposits in San Pablo Bay*, California, oral presentation at INTERCOH, Rio de Janeiro, Brazil (oral presentation).

2009   Bruce Jaffe,Theresa Fregoso, Amy Foxgrover, Mick van der Wegen, Dano Roelvink, , Neil Ganju, Kate Dallas, Patrick Barnard, Dan Hanes, John Chin, Don Woodrow, Mary McGann, Lynn Ingram, Shawn Higgins, Mark Marvin-DiPasquale and Elena Nielsen, *Morphological change of the San Francisco Estuary*, 90th Annual Meeting of the Pacific Division of the American Association for the Advancement of Science (AAAA), *San Francisco, CA* (oral presentation).

2009   Mick van der Wegen, Zheng Bing Wang, Huub Savenije and Dano Roelvink, *Process-based, long-term morphodynamic modelling to investigate conditions for equilibrium estuarine geometry*, oral presentation, EGU, Vienna, Austria (oral presentation)..

2008   Mick van der Wegen, Ali Dastgheib, Dano Roelvink, *Tidal inlet evolution along the Escoffier curve and empirical prism-cross section relationship using a 2D, process-based approach*, oral presentation at PECS, Liverpool, UK (oral presentation).

2008   Ali Dastgheib, Mick van der Wegen, Dano Roelvink, *Long-term morphodynamic modeling*, oral presentation, ICCE, Hamburg, Germany (oral presentation)..

2008   Mick van der Wegen, Bruce Jaffe, Dano Roelvink, *Decadal deposition in San Pablo Bay*, CalFed Science, Sacramento (oral presentation).

2008   Mick van der Wegen, Zheng Bing Wang, Ian Townend, Huub Savenije and Dano Roelvink, *Long-term morphodynamic evolution in an embayment using a numerical, process-based approach*, EGU, Vienna, Austria (poster presentation)

2008   Mick van der Wegen, Dano Roelvink, John de Ronde, Ad van der Spek, *Long-term morphodynamic evolution of the Western Scheldt estuary, the Netherlands, using a process based model*, COPEDEC VII conference, Dubai (oral presentation)

2008   Nguyen Thi Phuong Thao, Mick van der Wegen, Dano Roelvink, *Morphological behaviour of Nam Trieu Estuary, Vietnam*, oral presentation at COPEDEC VII conference, Dubai, (oral presentation)

2008   Mick van der Wegen, Bruce Jaffe, Dano Roelvink, *Application of a 2D numerical model in the San Francisco estuary to estimate morphodynamic change from global warming and sea level rise*, Ocean Science, Orlando, USA (poster)

2007   Mick van der Wegen, Zheng Bing Wang, Huub Savenije and Dano Roelvink *Morphodynamic Estuarine Evolution Using a 2D Numerical, Process-based Approach*, AGU fall meeting San Francisco, USA, (oral presentation)

2007   Mick van der Wegen, Zheng Bing Wang, Huub Savenije and Dano Roelvink, *Estuarine evolution using a numerical, process-based approach* proceedings RCEM 2007, Twente UT, Enschede, the Netherlands (oral presentation)

2007 Mick van der Wegen, Zheng Bing Wang, Huub Savenije and Dano
      Roelvink. *Energy dissipation in alluvial estuaries*, NCK days,
      IJmuiden, the Netherlands (oral presentation)
2006 Mick van der Wegen, Dang Quang Thanh and Dano Roelvink, *Bank
      erosion and morphodynamic evolution in alluvial estuaries using a
      process based 2D model,* proceedings ICHE conference 2006,
      Philadelphia, USA (oral presentation)
2006 Dano Roelvink, Giles Lesser and Mick van der Wegen, *Morphological
      modelling of the wet-dry interface at various timescales,* proceedings
      ICHE conference 2006, Philadelphia, USA (oral presentation)

About the author

Mick van der Wegen was born on December 31, 1971 in Amersfoort, the Netherlands. He attended the Johan van Oldenbarnevelt Gymnasium in Amersfoort from 1984 to 1990 and studied Civil Engineering at Delft University of Technology from 1990 to 1996. He graduated at the Department of Hydraulic Engineering in the Section Fluid Mechanics on the thesis "Turbulence in a shallow water mixing layer". The experimental work was carried out in the Laboratory for Fluid Mechanics at Delft University of Technology under guidance of dr. ir. J. Tukker, dr. ir. R. Booij and prof. dr. ir. J.A. Battjes. From 1994 to 1998 Mick studied part-time at the Royal Academy of Visual Arts in The Hague where he graduated in 1998.

In 1997, after his studies at Delft University of Technology, Mick van der Wegen joined the International Institute for Infrastructural, Hydraulic and Environmental Engineering (IHE) in Delft, where he worked in the core of Coastal Engineering and Port Development. First years were dedicated to teaching, MSc. study guidance, course management and capacity building projects. Main subjects of expertise became salt intrusion and density currents, integrated coastal zone management and morphodynamic modeling of coastal systems. From 2002 to 2004 he was member of the IHE Personnel Council.

In 2003 IHE transformed into UNESCO-IHE, the Institute for Water Education. From 2005 onwards Mick spent on average about 50% of his time on his PhD research. This work was partly financed by the internal UNESCO-IHE research fund and partly by USGS and CALFED. Within this latter framework, Mick (and his family) spent a beautiful spring from February to June 2008 in Santa Cruz, California USA, to work at the USGS office on morphodynamic hindcasting of San Pablo Bay.

Anneli and Mick live in Delft and have two daughters, Zide and Inez, and one son, Ben, who are respectively 8, 2 and 6 years old.